INTRODUCTION TO MUNICIPAL WATER QUALITY MANAGEMENT

INTRODUCTION TO MUNICIPAL WATER QUALITY MANAGEMENT

Johannes Haarhoff

Routledge
Taylor & Francis Group

LONDON AND NEW YORK

UNISA
university
of south africa
PRESS

First published 2023
by Routledge
4 Park Square, Milton Park, Abingdon, Oxon OX14 4RN

and by Routledge
605 Third Avenue, New York, NY 10158

Routledge is an imprint of the Taylor & Francis Group, an informa business

© 2023 University of South Africa

British Library Cataloguing-in-Publication Data
A catalogue record for this book is available from the British Library

ISBN: 9781032493954 (hbk)
ISBN: 9781032493961 (pbk)
ISBN: 9781003393573 (ebk)

DOI: 10.4324/9781003393573

Typeset in Times New Roman
by UNISA Press, South Africa

CONDENSED CONTENTS

ILLUSTRATIONS

TABLES

FIGURES

PREFACE

This book seeks to serve as a guide for chemists working at a municipality or a water treatment plant for the first time. It offers no introductory chapters on chemical reactions, stoichiometry, equilibria or kinetics – these are concepts already mastered by the chemist at undergraduate level. Instead, it aims to provide logical links between the chemist's specialist knowledge and the many seemingly empirical practices employed in the drinking water industry. During the first two decades of the 21st century, concerns about municipal water supply in South Africa focused mostly on the abundance and continuity of water supply, while attention may have been drawn away from other concerns about drinking water quality. Successful water supply hinges on both the quantity and quality of water – the domain of both engineer and chemist.

The foremost principle that informed the writing of the book is that it must be practical and quantitative. It is meant to empower the recently graduated chemist to translate theoretical concepts with confidence into quantitative solutions to practical problems. Where relevant, examples are included to demonstrate how to put theoretical concepts into practical use from first principles.

The first eight chapters are devoted to topics narrowly focused on the monitoring and treatment of drinking water. The final four chapters take a broader view of municipal water supply, dealing with recreational water and bulk water supply. Although water supply authorities will probably use specialists in these areas for implementation, it is likely that the municipal chemist will be drawn upon to liaise with these specialists. A brief introduction to the main concerns and general principles is therefore deemed to be appropriate.

The drinking water industry is a well-regulated professional environment with a plethora of regulations, standards and guidelines. A young chemist entering this field may perceive it as being an exclusively rule-based environment, missing the point that it is solidly underpinned by logical thinking and mostly solid science. If this book encourages the practitioner to find and appreciate the scientific reasons beyond the rules, it served its purpose.

FOREWORD

For many years, one yearned for a book that brings together years of accumulated knowledge generated on many platforms of water research, particularly associated with the quality of drinking water and the practical experience gained from drinking water treatment plants. This is the confluence of academia and industry rivers of knowledge trajectories geared towards empowering postgraduate students from various disciplines – Chemistry, Environmental Engineering, Biotechnology – who are enrolled for postgraduate studies to undertake research projects on topics centred around drinking water treatment. This is a book whose genesis is research and it also sharpens one's knowledge towards problem solving as it introduces the vital principles of water treatment which would not normally be found in a typical textbook. Professor Johannes Haarhoff, the author of this book, spent ten years in engineering practice before moving into academia for a further 30 years, teaching undergraduate programmes at university, supervising master's and doctoral students and acting as specialist consultant to various water treatment plants and municipalities. Professor Haarhoff writes this splendid book as arguably his final contribution to water research after spending the latter years of his outstanding career history at the Nanotechnology and Water Sustainability (NanoWS) Research Unit, College of Science Engineering and Technology (CSET), University of South Africa (Unisa), where he had been employed as a research professor.

Prof Bhekie B Mamba, PhD, CChem, CSci, FRSC, FWISA
Executive Dean: College of Science Engineering and Technology (CSET)
University of South Africa

PART I

DRINKING WATER QUALITY

PART 1

STANDARDS

1.1. INTRODUCTION

To the public, drinking water quality is mostly a simple question of "Is it safe to drink?" and then expecting a simple "yes" or "no". An experienced chemist appreciates that this question is more complex which warrants an extended, qualified answer. Poor communication between water professionals and the general public lies at the heart of the doubts often harboured against their water suppliers and provides fertile breeding ground for public scepticism and fear. Proper understanding, interpretation and communication regarding water quality and its standards are key prerequisites for chemists working in this field.

1.2. CATEGORIES OF WATER QUALITY PARAMETERS

Hundreds of standardised tests are available for the testing of drinking water; each testing for a different parameter. For meaningful discussion, it is useful to group these parameters – either by their potential risks, or by analytical considerations.

1.2.1. Water Quality Parameters Grouped by Risk

There are four major risks guiding the establishment of water quality standards:

- Aesthetic risks must be avoided to give the consumer confidence in the quality of drinking water. From time immemorial, survival of the human species depended, among others, on the ability of individuals to avoid water of poor quality. A keen natural sense developed to recognise some danger signs, inherited by us all – everybody is likely to refuse very murky water, or water with a greenish tinge or pungent odour. These adverse effects are analytically measured, for example, as turbidity, taste, odour, colour and temperature. Another concern arises when dissolved species precipitate during the normal water use cycle, such as iron (Fe), manganese (Mn) and copper (Cu). The Fe and Mn precipitates impart a brownish-black stain to clothing, linen and other surfaces; Cu leaves green stains on enamelware.

- Acute health risks pose the danger that health complications may follow within minutes or hours after ingestion of the water. These effects could be caused by chemical compounds (such as arsenic, cyanide or very high levels of fluoride) or microbiological organisms (such as food poisoning by viruses or bacteria).

- Chronic health risks pose health complications which are only evident after years or decades of continuous exposure. The effects could be manifested as cancers, damage to the nervous system, respiratory failure, radiation effects, etc.

- Operational risks are associated with failures in treatment plant operation. Risk management requires control parameters that are quickly and routinely measured to ensure consistently good water quality. Not only must the aesthetic, acute and chronic health risks be avoided, but the treated water must also be non-corrosive. Corrosive water may cause irreversible damage to pipes, valves, pumps and tanks in the water distribution system, leading in turn to leaks, supply interruptions or expensive replacement. Corrosion therefore poses an economic risk. It is a complex phenomenon caused by the combined effects of pH, alkalinity and temperature – to be unpacked in Chapter 5.

1.2.2. Water Quality Parameters Grouped by Analytical Considerations

Analytical laboratories prefer to categorise water quality parameters in relation to the methods and expertise required for their analysis. The typical grouping is:

- Microbiological determinands. These include viruses, protozoan parasites and bacteria.
- Physical and aesthetic determinands. These include chlorine species, colour, turbidity, etc.
- Chemical macro-determinands. These chemical species are tolerated at higher levels than the micro-determinands and the limiting concentrations are expressed in mg/L. Examples are sulphate, chloride and sodium.
- Chemical micro-determinands. Their limiting concentrations are lower and expressed in μg/L. Examples are cadmium, arsenic and cyanide.
- Organic determinands. These include total organic carbon, trihalomethanes and phenols.

1.2.3. Mapping of Water Quality Parameters

It is beyond the scope of this document to discuss the specific risks for all the regulated water quality parameters in detail. In South Africa, the water quality regulations are published by the South African Bureau of Standards (SABS) and updated every few years. These regulations will be extensively referenced throughout Chapters 1 and 2 as SANS (South African National Standard) 241-1 and SANS 241-2; so listed in the references at the end of this chapter. An abbreviated summary of risks from SANS 241-1 is provided in Tables 1.1 to 1.5.

Table 1.1   Microbiological water quality parameters[a]

Determinand	Risk Category	Specific Effect
E. coli or faecal coliforms	Acute health	Faecal pollution indicator
Cytopathogenic viruses	Acute health	Damage to cells
Protozoan parasites	Acute health	Gastro-enteritis, diarrhoea
Total coliforms	Operational	Faecal pollution indicator
Heterotrophic plate count	Operational	Microbiological activity indicator
Somatic coliphages	Operational	Faecal pollution indicator

[a] Taken from Table 1 of SANS 241-1 (2015)

Table 1.2   Physical and aesthetic water quality parameters[a]

Determinand	Risk Category	Specific Effect
Free chlorine	Chronic health	Disinfectant
Monochloramine	Chronic health	Disinfectant
Colour	Aesthetic	Consumer resistance
Conductivity at 25°C	Aesthetic	Dissolved salts indicator
Odour or taste	Aesthetic	Consumer resistance
Total dissolved solids	Aesthetic	Unpleasant taste
Turbidity	Operational	Particle removal indicator
Turbidity	Aesthetic	Consumer resistance
pH at 25°C	Operational	Corrosivity, taste, dissolved metals

[a] Taken from Table 2 of SANS 241-1 (2015)

Table 1.3   Chemical (macro-determinands) water quality parameters[a]

Determinand	Risk Category	Specific Effect
Nitrate as N	Acute health	Methemoglobinemia
Nitrite as N	Acute health	Methemoglobinemia
Sulphate as SO_4^-	Acute health	Diarrhoea
Sulphate as SO_4^-	Aesthetic	Bitter, salty taste
Fluoride as F^-	Chronic health	Tooth enamel, skeletal fluorosis
Ammonia as N	Aesthetic	Taste and odour
Chloride as Cl^-	Aesthetic	Salty taste
Sodium as Na	Aesthetic	Taste, hypertension
Zinc as Zn	Aesthetic	Taste, milky appearance

[a] Taken from Table 2 of SANS 241-1 (2015)

Table 1.4   Chemical (micro-determinands) water quality parameters[a]

Determinand	Risk Category	Specific Effect
Antimony as Sb	Chronic health	Diarrhoea, liver damage
Arsenic as As	Chronic health	Skin lesions, skin cancer
Cadmium as Cd	Chronic health	Kidney damage
Total chromium as Cr	Chronic health	Gastro-intestinal cancer
Cobalt as Co	Chronic health	Heart damage, thyroid damage
Copper as Cu	Chronic health	Taste, staining, gastro-intestinal
Cyanide as CN^-	Acute health	Nervous system, thyroid
Iron as Fe	Chronic health	Fatigue, joint pain
Iron as Fe	Aesthetic	Taste, stains, deposits
Lead as Pb	Chronic health	Neurological damage

Manganese as Mn	Chronic health	Neurological damage
Manganese as Mn	Aesthetic	Taste, staining
Mercury as Hg	Chronic health	Damage to nervous system, liver
Nickel as Ni	Chronic health	Skin irritation
Selenium as Se	Chronic health	Liver damage, slow growth of hair, nails
Uranium as U	Chronic health	Radioactivity
Vanadium as V	Chronic health	Slow growth, respiratory symptoms
Aluminium as Al	Operational	Possible neurotoxic effects

[a] Taken from Table 2 of SANS 241-1 (2015)

Table 1.5   Organic water quality parameters[a]

Determinand	Risk Category	Specific Effect
Total organic carbon as C	Chronic health	Indicator of organic pollution
Chloroform	Chronic health	Low risk of cancer
Bromoform	Chronic health	Low risk of cancer
Dibromochloromethane	Chronic health	Low risk of cancer
Bromodichloromethane	Chronic health	Low risk of cancer
Microcystin as LR	Chronic health	Skin irritation
Phenols	Aesthetic	Odour

[a] Taken from Table 2 of SANS 241-1 (2015)

1.3. NATIONAL AND INTERNATIONAL WATER QUALITY GUIDELINES

There are numerous quality guidelines for potable water in the world. The World Health Organisation (WHO) provides one of the most authoritative which are revised every few years. The European Union (EU) provides a regional EU standard. Many countries have adopted their own standards, such as South Africa's SANS 241. The United States Environmental Protection Agency (USEPA) provides the most comprehensive country-based drinking water standard. Specific water suppliers go even further and derive their own quality guidelines which they use as an internal limit to guide their operation. Rand Water, for example, applies such a standard.

It is surprising that there are large differences among these standards – should they not all be identical? Quantitative comparisons had been made by numerous researchers to illustrate the point. Mamba et al. (2008), for example, compared the health-related parameters of four standards – the WHO guideline, the regional EU standard and the national standards of the Netherlands and South Africa. Many discrepancies were noted — some parameters are regulated by some, while ignored by others. In some cases, SANS 241 was the lowest; in other cases, the highest. A general pattern was observed, namely that the Dutch standard was generally the strictest, followed by the EU standard, then the WHO standard, with SANS the most lenient. SANS covered the chemical health-related determinands fairly well (only one out of 21 determinands considered by the total group was not covered – bromate) but has much poorer coverage of organic determinands (only two covered of the nine determinands considered by the total group), namely, total trihalomethanes and total organic carbon.

There are good reasons to justify most of the differences. In some regions, specific problems are prevalent which require greater emphasis on those compounds. In the USA, for example, radioactive radon gas is associated with groundwater in some regions, putting radon in their list while absent in most other regions. Arsenic is prevalent in the shallow wells in Bangladesh, making it a priority determinand in those parts. The standards are also dictated by practical and economic considerations. Large rural areas in Namibia, for example, rely on groundwater exclusively which many might find unacceptably brackish. The Namibian limit for total dissolved salts, unsurprisingly, is therefore higher than the South African limit. In Europe, as a final example, a maximum temperature (usually 15°C) is included in some guidelines as an aesthetic parameter — most people prefer cooler water. In South Africa, this limit would be ridiculous as our surface water temperature in summer frequently exceeds 20°C. To cool water artificially at municipal scale is completely impractical and uneconomical.

The numerical limits provided by the water quality standards should be interpreted with circumspection. Later in this chapter, it will be demonstrated that the setting of these limits is anything but an exact science. A standard of say 100 units/L certainly does not mean that water with 90 units/L is perfectly safe, and that 110 units/L is completely unfit for drinking! Water suppliers in the Netherlands, for example, strive to continuously improve the water quality with respect to all parameters within the usual constraints of cost and practicality. It does not matter how far or close the parameter may be from the standard. A water supplier must be on the lookout for any quality parameter that creeps upwards, however slowly, even if it is still way below the limit. Act proactively to take precautionary actions — procrastination until the limit is finally exceeded may be too late.

1.4. HOW ARE WATER QUALITY PARAMETERS SELECTED?

The establishment of a water quality standard is a two-step process. In the first, covered in this section, the parameters are selected. In the second, covered in Section 1.5, the numerical limits for each parameter are established.

There is no fixed algorithm for selecting the potential candidates for regulation, and how to screen them for eventual inclusion in the final list of regulated compounds. Professionals from many disciplines will raise their concern if they suspect some systematic problem that may be water related. Medical practitioners might observe some patterns in the complaints from their patients. Furthermore, water suppliers might observe some change in the nature or frequency of complaints; farmers might start using new chemicals in large quantities; pharmaceutical companies might introduce large quantities of new products that might find their way into our water sources; biologists might find trends towards extinction of some plant and animal species — the list continues. In this way, some consensus emerges that some new compounds might warrant more detailed attention. This is usually followed by a period of more systematic and intensive monitoring to see how much of the contaminant is actually present in the water environment. At the same time, others will assemble the physical and chemical characteristics of the contaminant, its pathways through the environment and the human body, the possibility of removing the contaminant during treatment, etc. After some more time, a multidisciplinary group of experts reviews the available evidence and, if warranted, the parameter will become a candidate for regulation. This process, greatly simplified here, usually runs over a period of a few years to a decade or two. The potential danger of trihalomethanes, for example, was raised for the first time in 1976, but it took almost 15 more years of systematic analysis before enough was known to add trihalomethanes to the regulated list of water parameters.

In South Africa, the SANS standard is regularly updated by an expert committee invited by the SABS, with the latest standard at the time of writing released in 2015.

1.5. HOW ARE WATER QUALITY LIMITS DETERMINED?

A typical example of how a water quality standard is developed is provided by a public health statement issued on cobalt in drinking water (Agency for Toxic Substances and Disease of Registry 2004).

The quantitative setting of water quality limits is only possible once we have answers to the following questions:

- What is cobalt? (A good understanding of cobalt — where it is found, in which chemical compounds and valencies it may be found, their reactivities, physical and chemical properties, etc.)
- What happens to cobalt when it enters the environment? (The pathways into the environment by natural and anthropogenic activities, with its subsequent flow and fate.)
- How might humans be exposed to cobalt? (Estimates of the concentration of cobalt in the three major modes of exposure — air inhalation, intake of food and water, exposure to skin.)
- How does cobalt enter and leave the human body? (A knowledge of the metabolic pathways through the body — air in lungs, absorption in blood, where it may be concentrated, how it is excreted.)
- How can cobalt affect my health? (Awareness of possible health effects — acute, short-term exposure and long-term exposure.)

While the logic of the approach may seem obvious, the quantification of the many intermediate steps is rife with assumptions, supported by often noisy or incomplete experimental evidence. The estimate of the mass released into the environment can be made with reasonable precision, but the degradation and metabolic pathways are more difficult. Environmental monitoring can check these estimates. The toxicological work is even more complex. Rats, mice, rabbits or other species are exposed to different levels of the contaminants for different periods to produce different measures such as LD50 (Lethal Dose which kills 50%), NOAEL (No Observed Adverse-Effect Level) and LOAEL (Lowest Observed Adverse-Effect Level). Expressed on a (mass of contaminant) / (body mass of test animal) basis, it allows scaling up to human level by professional toxicologists.

Example 1.1

This example shows how a 20 µg/L limit for antimony was set for drinking water, based on a publication by the World Health Organisation (WHO) (2003). A critical comparison of experimental animal studies suggested a NOAEL of 6 mg/kg (mass of antimony per kg of body weight per day) based on a study with rats. Above this concentration, the rats suffered a decrease in weight gain and reduced food and water intake.

For extrapolation to humans the NOAEL safety factor = 1 000

Total daily intake (TDI) = 6 000/1 000 = 6 µg/kg

Humans derive about 10% of their TDI of antimony from drinking water,

therefore TDI from water = 6 x 0.10 = 0.6 µg/kg

A human of 60kg drinks 2L of water per day, therefore

Allowable concentration of antimony = 0.6 x 60/2 = 18 say 20 µg/L

1.6. THE CHALLENGE OF EMERGING CONTAMINANTS

1.6.1. Classification of emerging xenobiotic organic compounds

Xenobiotic compounds (compounds that are foreign to living organisms) are proliferating in the modern industrial era. The new compounds have grown in number to the point where we now need to identify classes or categories of compounds to allow meaningful discussion. At present, the Chemical Abstracts Service (CAS) lists 60 million registered organic and inorganic substances in the world (available at http://www.cas.org/). The USEPA considers 84 000 of these to be commercially important chemicals. Remember that SANS 241 includes less than 50 contaminants, with all determinands included!

Persistent Organic Pollutants (POPs) form a small, clearly defined group of organic compounds that resist photolytic, chemical and biological degradation and have known toxic properties. They have been studied relatively well during the past 20 years. The Stockholm Convention is an international agreement designed to eventually rid the planet of these harmful compounds. It started by listing 12 compounds (informally labelled as the "dirty dozen") and has since grown to 23 different POPs which are firmly recognised, with a further six being reviewed for possible recognition later.

Endocrine Disrupting Compounds (EDCs) are compounds that are suspected of possibly interfering with the functioning of natural hormones in living organisms. This interference manifests in different ways with a variety of symptoms not covered here. From a human perspective, endocrine disruption is the main concern about xenobiotic compounds.

Pharmaceuticals and Personal Care Products (PPCPs), organic in nature, emanate from the growing consumer use of compounds present in medication, household chemicals and toiletries.

It is difficult to identify the compounds of concern from all the new compounds that are continuously registered. The precautionary principle dictates that all new compounds should be treated as potentially harmful until better scientific evidence is available for accepting or rejecting the hypothesis. The compilation and maintenance of lists of CECs and EDCs are therefore important starting points for systematic evaluation. A number of these lists are freely available in the form of databases with valuable information listed for each compound, such as its physical properties, toxicity, reference to published studies, regulatory status, its CAS registered number, appropriate analytical methods, etc. The most prominent databases for the contaminants of emerging concern are:

- The TEDX (The Endocrine Disrupting Exchange) database, with more than 1000 compounds. Maintained by TEDX, available at http://endocrinedisruption.org/endocrine-disruption/tedx-list-of-potential-endocrine-disruptors/overview.
- The IRIS (Integrated Risk Information System) database, with more than 550 compounds. Maintained by the USEPA, available at http://cfpub.epa.gov/ncea/iris/index.cfm?fuseaction=iris.showSubstanceList.
- The SIN (Substitute It Now) list with 406 compounds. Maintained by the International Chemical Secretariat (ChemSec), available at http://www.chemsec.org/what-we-do/sin-list.
- The HSDB (Hazardous Substances Data Bank) database, with information on 5 756 compounds. Maintained by the US Department of Health and Human Services, available at http://sis.nlm.nih.gov/enviro/hsdbchemicalslist.html.
- Country lists. Many countries have drawn up their own lists of CECs. A reference to some of these lists is available at http://ec.europa.eu/environment/archives/docum/pdf/bkh_annex_02_03.pdf.
- The Household Products database with information and ingredients on 14 000 consumer brands. Maintained by the US Department of Health and Human Services, available at http://hpd.nlm.nih.gov/.

The classes identified are not mutually exclusive, complicating systematic consideration. The simplified graphic in Figure 1.1 shows that one compound may belong to more than one class, and that not all PPCPs are CECs.

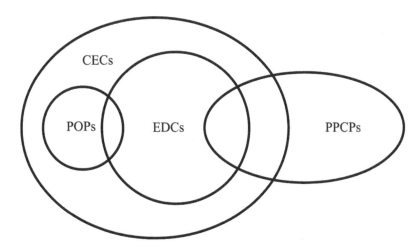

Figure 1.1   Relationships among classes of emerging contaminants

The Water Research Commission of South Africa recently considered it timely to present a consolidated view of past and ongoing research to a national workshop on POPs and CECs, in order to map out a strategy for guiding future research into these important areas (Coetzee et al., 2016). A number of the findings are used to close this chapter.

1.6.2. Sampling Considerations

The sheer magnitude of compounds and the complexity of their analytical measurement call for preliminary screening steps to reduce the sampling sites and the required monitoring parameters, to render monitoring tractable and affordable. The first screen is to focus on sites where CECs are most likely to be found, such as harbours, polluted rivers, areas where pesticides are applied, cattle feedlots, streams downstream of informal settlements, and urban catchments. Second, bioassays must be used to test whether there was actual EDC activity at the selected sampling sites. Only if such activity is observed, a third tier of more intense monitoring and analytical analysis is warranted.

CECs are present in water, sediments and in water-based organisms. Each of these sample types have strengths and weaknesses. The transport of water through rivers, streams and small impoundments is fairly quick and short-lived contaminant peaks may escape detection unless monitoring frequencies are high. The low water solubility of many of the CECs requires very low detection limits if water samples are analysed. A short-lived contaminant peak washed rapidly through an impoundment will, however, leave its trace for much longer in the bottom sediments, where many of the CECs tend to adsorb. To detect locations where frequent contaminant peaks occurred, or where contaminants were released in low but persistent concentration, the sampling of sediments may be a more appropriate sample type. Living organisms also tend to accumulate many of the CECs in their tissue and may be better indicators of long-term contaminant exposure than the water itself.

The first step towards a systematic "sweep" of South Africa for CECs is to work at catchment level. A proper audit of economic activities in each catchment area may point to those CECs likely to be a problem and which should be sampled for – an input-directed approach. The second step would be resource-directed namely to start sampling and testing the water resource according to a tiered approach. Initial tests should be restricted to bioassay screening for EDC activity, only followed by multi-parameter chemical analysis if a possible hotspot is suggested.

1.6.3. Analytical Considerations

The costs and difficulties of analytical testing have been a consistent frustration to researchers since the inception of CEC research. While South Africa certainly has the professional ability and sophisticated instrumentation to perform the analyses required, the cost of EDC testing is prohibitive, putting such testing beyond the financial reach of most water supply authorities. Worse, some services may be not commercially available at all. The only way to drive down costs and to encourage the investment required for readily affordable, available commercial analyses is to ensure a steady supply of samples for a guaranteed period of five to ten years. With this scenario, private and research laboratories might be encouraged to invest into the necessary expertise and equipment. The analytical workload should be ideally shared between state-run laboratories, local commercial laboratories and international laboratory services providers to get the work done at the lowest price.

1.6.4. The Water Use Cycle

The complexity of dealing with the multi-dimensional problem of CECs is best illustrated by an overall view of the water use cycle, depicted in Figure 1.2.

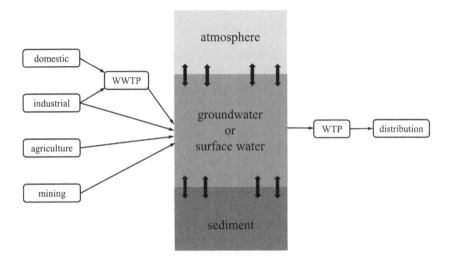

Figure 1.2   Pathways of contaminants in the water use cycle

The main entry points of POPs and CECs are the runoff into rivers and impoundments, or seepage into groundwater. The inputs considered are grouped by domestic, industrial, agricultural or mining origin. The domestic and industrial (partly) inputs are routed through wastewater treatment systems before they reach the surface water or groundwater, while the agricultural and mining inputs are discharged directly.

Once the CECs enter the water resource, a complex partitioning process takes place between the solid phase (sediments or geological stratum), liquid phase (free or interstitial water) and gas phase (the atmosphere for impoundments and rivers, or vadose zone for groundwater). The drinking water value chain continues from the liquid phase, which makes the concentration of CECs remaining in the water phase of primary concern.

The raw water abstracted from the water resource is routed through a water treatment plant, where some of the raw water contaminants may be removed. Nevertheless, treatment chemicals are also added, which must be considered for possible CECs. Furthermore, the treated water spends some time, up to a week, in a distribution system where the formation of disinfection by-products, or the leaching from the materials used for pipes, valves, water meters, pumps and reservoirs may compromise the quality of the water eventually delivered to the taps of consumers.

A part of the South African population in rural areas, or on the urban fringes, must contend with a domestic water use cycle less complete than depicted in Figure 1.2. There may not be any formal treatment when people take their drinking water directly from a river, well or impoundment. More commonly, there is only a partial distribution system which does not extend into private homes, when people carry water in buckets from central standpipes to their homes. In the latter case, there are many potential contaminant entry points along this "human pipe", such as inappropriate materials used for buckets and plastic cans, biofilms on the inside of poorly cleaned containers, recontamination of water in the home, etc.

1.7. REFERENCES

Agency for Toxic Substances and Disease Registry. 2004. *Public Health Statement on Cobalt*. http://www.atsdr.cdc.gov/phs/phs.asp?id=371&tid=64.

Coetzee, L.Z., Gumbi, N., Haarhoff, J., Mamba, B.B., Msagati, T.A.M., Nkambule, T., and Wanda, E.M.M. 2015. Status Quo Report on the State of Knowledge on Persistent Organic Pollutants and Contaminants of Emerging Concern. Unpublished WRC Report 1087. Pretoria: Water Research Commission.

Mamba, B.B., Rietveld, L.C., and Verberk, J.Q.J.C. 2008. "SA Drinking Water Standards under the Microscope." *Water Wheel*, 7(1): 24–27. Pretoria: Water Research Commission. http://www.wrc.org.za/wp-content/uploads/mdocs/WaterWheel_2008_01_WW%20Jan-Feb%2008.pdf.

South African Bureau of Standards. 2015. South African National Standard (SANS) 241-1. *Drinking Water Part 1: Microbiological, Physical, Aesthetic and Chemical Determinands*. Pretoria: South African Bureau of Standards.

World Health Organisation. 2003. *Antimony in drinking-water*. Report WHO/SDE/WSH/03.04/74. Geneva: World Health Organisation. http://www.who.int/water_sanitation_health/dwq/chemicals/antimony.pdf.

COMPLIANCE

2.1. INTRODUCTION

The drinking water standard adopted by a drinking water supplier (the latest version of SANS 241 applies in South Africa) provides the quantitative foundation for water quality management. Having a water quality standard, however, is but the first step towards ensuring that the public is continuously provided with safe drinking water. It may be simple to apply the standard to a single water sample, but how does one apply the standard to a large, complex distribution network supplying thousands of customers? This is essentially a statistical problem – how often, where and how many samples should be taken and tested to adequately ensure that the public gets drinking water compliant to the water quality standard? Therefore, it now falls upon the water quality manager, necessarily someone with a strong scientific background, to design a monitoring programme to meet this important need.

SANS 241 does not only provide the limiting concentrations of the different parameters but helpfully provides explicit guidelines to assist. To the uninitiated, they may seem nothing more than a plethora of dos and don'ts, but the sampling guidelines are underpinned by solid scientific reasoning. This chapter systematically unpacks the SANS 241-2 guidelines, with examples to demonstrate their application.

2.2. THE HACCP APPROACH

The Hazard Analysis and Critical Control Points (HACCP) concept was originally developed in the food industry and defined as "a preventative food safety system in which every step in the manufacture, storage and distribution of a food product is scientifically analysed for microbiological, physical and chemical hazards" (Food and Agriculture Organization 2001). A much wider audience, including the drinking water sector, has since embraced HACCP's robust framework for risk analysis and mitigation. It hinges on seven principles enumerated here with comments on how they are addressed by South African drinking water treatment practice:

- Identify and analyse hazards: SANS 241-1 is based on a list of 43 determinands that might pose risks to consumers, discussed in Chapter 1.

- Determine the critical control points (CCPs): SANS 241-2 provides clear guidelines on the number and position of the CCPs, discussed in Section 2.3.

- Establish critical limits for the quality at each CCP: SANS 241-1 defines concentration limits for the 43 regulated determinands, discussed in Chapter 1.

- Establish a monitoring procedure: SANS 241-2 provides monitoring procedures for compliance with its regulations, detection of operational errors and verification of corrective actions, discussed in Sections 2.4 to 2.6.

- Establish corrective action: SANS calls for corrective action where necessary, without being specific.

- Verify the HACCP plan: As verification, SANS 241 is reviewed roughly every five years by an expert panel to ensure best practice.

- Keep record: Beyond the scope of SANS 241 and not within the scope of this text.

It is evident that SANS 241 covers most of the required HACCP steps. The remaining part of the chapter will discuss its 2015 guidelines in general terms. For the design of actual monitoring systems, it is essential to consult the more detailed, latest version of SANS 241.

2.3. CHOOSING CRITICAL CONTROL POINTS (CCPs)

Water supply authorities must deal with three types of water — raw water, treated water and water at the point of use. CCPs are required for all three types. For each raw water source, there must be a CCP; for each treatment plant there must be a CCP after treatment. For each distribution zone, there must be one or more CCPs. Some water suppliers (water boards, for example) sell treated water in bulk to other suppliers (municipalities, for example) – these suppliers obviously need to include only the water types relevant to their operation.

For water supplied through a reticulation network, the selection of CCPs is not so simple. There may be an erroneous perception that water quality remains unaltered after treatment. There are numerous drivers of water quality changes after treatment:

- It takes about four to eight hours for water to flow through a typical treatment plant. In contrast, the treated water spends days (even two weeks in extreme cases) within the reticulation network system before used by consumers. The much longer time in the reticulation network allows slower chemical and microbiological processes to have an adverse effect on water quality.

- The long residence time in the reticulation network complicates chemical disinfection. Disinfectants decay or react over time. To ensure that disinfection residuals are maintained up the point of consumption, monitoring within the reticulation network is required.

- Open water surfaces provide breeding ground for unwanted larvae and other organisms. If the screens of roof ventilators on reservoir roofs, for example, are not well maintained, the water quality at the consumer could be compromised.

- Slow sedimentation of particles in reservoirs is inevitable. A thick layer of bottom sludge could be disastrous if a slug of sludge is drawn from the reservoir at low water level. Reservoirs should be inspected, drained down and cleaned when necessary, normally every few years.

- Similarly, some sedimentation will occur in pipes, especially where low flow velocities are encountered towards the far ends of reticulation networks. A flushing programme via fire hydrants may be required from time to time to prevent these accumulations from impacting on consumers.

- Whenever there is a pipe failure, there is the danger of contamination of the water or ingress of foreign matter. Good practice suggests that thorough flushing and in-situ disinfection should follow all pipe repairs. If these preventative steps are bypassed, water quality will be impacted.

It follows that it is more difficult to maintain good water quality in large distribution zones where the pipes are longer and more, with a resulting increase in hydraulic retention time. The selection of CCPs in the reticulation network requires careful judgment. SANS 241-2 stipulates that:

- Each independent distribution zone should have its own CCPs.
- All outlets of reservoirs and water towers are considered to be CCPs.

- The furthest points of all distribution zones are considered to be CCPs.
- At least 80% of each distribution zone should be covered by sampling.
- Areas with elevated risks such as hospitals, schools or network pipes with dead ends must be included as CCPs.

2.4. MONITORING FOR WATER QUALITY RISK

The first priority is to identify those determinands that pose major water quality risks, if any. SANS 241-2 therefore stipulates a full analysis of all the regulated determinands at all the CCPs once a year, taken at a time when it is anticipated that the quality will be at its worst. For each determinand, the results could take any of the forms indicated in Figure 2.1.

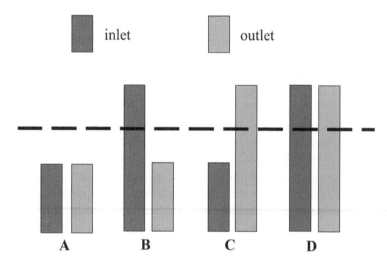

Figure 2.1   Possible outcomes of risk monitoring

Consider the four possible outcomes depicted in Figure 2.1:

- Case A is the most favourable, where neither the raw nor the final water exceeds the SANS 241 limit. In this case, no more monitoring for this determinand is required until the next sampling in the following year, unless it is necessary for operational monitoring (Section 2.5).
- Case B indicates that the raw water exceeds the standard, but that treatment is efficient for removing it to below the acceptable limit. However, any small mishap in operation may push it over the limit – therefore, operational monitoring must be stepped up (Section 2.5)
- Case C shows that a determinand increased during treatment, pointing to contamination by treatment chemicals or formation of disinfection by-products. Aluminium, iron, disinfectant residuals, ammonia and trihalomethanes are typical examples relating to operational processes and the type of treatment chemicals used. This calls for a review of the treatment strategy and the selection of treatment chemicals.

- Case D is the most unfavourable, where both raw and treated water exceed the limit. This calls for an engineering intervention to provide better or more treatment processes. Monitoring must be stepped up until the problem is solved (Section 2.5).

2.5. MINIMUM MONITORING FOR OPERATIONAL RISKS

Following the annual sampling for compliance risks, the next step is to design a monitoring programme to ensure that the treatment plant and reticulation networks are operated properly. As a minimum, there is a core set of parameters to be measured as shown in Table 2.1.

Table 2.1. Minimum monitoring for process indicators[a]

	Raw Water CCPs	Treated Water CCPs	Distribution[b]
Conductivity or TDS	Daily	Daily	Not applicable
pH value	Daily	Once per 8h shift	Fortnightly
Turbidity	Daily	Once per 8h shift	Fortnightly
Disinfectant residuals	Not applicable	Once per 8h shift	Fortnightly
E. coli or faecal coliforms	Not applicable	Weekly	Fortnightly
Heterotrophic plate count	Not applicable	Weekly	Fortnightly
Treatment chemicals	Not applicable	Monthly	Not applicable

[a] Taken from Table 1 of SANS 241-2 (2015). For groundwater sources, monthly samples are adequate.
[b] At all CCPs *E. coli* or faecal coliforms at additional points to meet Table 2.2.

E. coli or faecal coliforms indicate potential acute health risks and therefore are subject to more specific requirements. All the CCPs in the reticulation networks are automatically sampling points for *E. coli* and faecal coliforms. In addition, there must be a minimum number of samples based on the population served by that particular distribution zone, shown in Table 2.2. If this number is more than the CCPs, then additional sampling points must be established, or the CCPs must be sampled more regularly to get to the required number.

Table 2.2. Minimum samples for *E. coli* or faecal coliforms in distribution systems[a]

Population Served	Minimum Number of Samples Per Month
Less than 5 000	2
5 000 to 100 000	1 per 5 000 head of population +1
100 000 to 500 000	1 per 10 000 head of population +11
More than 500 000	1 per 20 000 head of population +36

[a] Taken from Table 2 from SANS 241-2 (2015)

For groundwater, where quality variations are slower and generally less severe, the monitoring frequency can be reduced to once per month provided that the limits are not exceeded. If they are exceeded, the frequency must be stepped up to that stipulated in Table 2.1.

If the minimum monitoring of Table 2.1 detects one or more samples that do not meet the standard, corrective action must be taken and the sampling frequency must be increased and continued until the problem is resolved satisfactorily.

Example 2.1

A town with 30 000 inhabitants gets it raw water from a dam, as well as a borehole. The borehole water is chlorinated and then blended with the treated dam water. The combined stream is then supplied to two distribution zones, each with own reservoir. Zone A supplies 10 000 people, while Zone B supplies 20 000 people. If everything goes well (no exceedances), estimate the annual number of a) turbidity and b) *E. coli* samples to be measured if there are, on average, two shifts per day.

Number of raw water CCPs = 2 (one for the borehole water before chlorination, one for dam water)

Number of treated water CCPs = 2 (one for the chlorinated borehole, one for treated dam water)

Number of Zone A CCPs = 2 (one after reservoir, one at furthest end, after study of layout)

Number of Zone B CCPs = 3 (one after reservoir, two at opposite furthest ends, after study of layout)

Zone A requires 10 000/5 000 + 1 = 3 sampling points, therefore 1 additional to CCPs

Zone B requires 20 000/5 000 + 1 = 5 sampling points, therefore 2 additional to CCPs

Turbidity = 365 (raw dam water) + 12 (raw borehole water) + 2x365 (treated dam water) + 12 (chlorinated borehole water) + 26x5 (five distribution CCPs) = 1 237 samples/year

E. coli = 52 (treated dam water) + 12 (chlorinated dam water) + 26x8 (five CCPs and three additional points) = 272 samples/year

---

2.6. CLOSER MONITORING OF RISK DETERMINANDS

The annual sampling survey of the CCPs identifies those determinands that pose a potential risk — a process discussed in Section 2.4. These determinands must be incorporated into the ongoing monitoring strategy in accordance with Table 2.3, which shows that the frequency depends on the type of risk, a matter discussed in Chapter 1. An acute health risk requires more urgency than a chronic health risk which explains why it requires more frequent sampling. The sampling points to be included depend on the nature of their non-compliance:

- If only the raw water is non-compliant, then the raw and treated CCPs of that source must be sampled.
- If the final water is non-compliant, then the raw, final and all subsequent distribution CCPs must be sampled.
- If only the distributed water is non-compliant, then only the distribution CCPs must be sampled.
- The higher frequency of monitoring should be kept up until it no longer constitutes an unacceptable risk.

Table 2.3: Sampling frequencies for risk determinands[a]

	Raw Water	Final Water	Critical Control Point
Acute health chemical risk	weekly	weekly	monthly
Chronic health risk	monthly	monthly	monthly
Aesthetic risk	monthly	monthly	quarterly
Operational risk	weekly	weekly	monthly

[a] Extract from Table 3 of SANS241-2 (2015)

Example 2.2

The annual sampling of the town described in Example 2.1 revealed that three determinands exceeded the standard, namely, ammonia in the raw dam water, cadmium in the raw and final borehole water and colour in the distributed water of Zone A. If we assume that it may take a year to resolve the underlying problems, estimate the number of samples for each of these determinands for a year.

Ammonia poses an aesthetic risk; therefore monthly samples need to be drawn in raw dam water only.

Number of ammonia samples = 12 (raw dam water)

Cadmium poses a chronic health risk; therefore, monthly samples need to be drawn at all CCPs.

Number of cadmium samples = 12x2 (raw and chlorinated borehole water) + 12x5 (distribution CCPs) = 84 samples

Colour poses an aesthetic risk; therefore, monthly samples need to be drawn at distribution CCPs only.

Number of colour samples = 12x5 (distribution CCPs) = 60 samples.

2.7. WATER QUALITY INDICES IN COMPLIANCE WITH SANS 241-2

The prescribed monitoring of SANS 241, even if there were no quality problems whatsoever, produces thousands of data points every year which are clearly very tedious to interpret and completely unsuited for communication to the public or higher levels of management. Some form of data reduction or summary is required. In order to prevent selective reporting or "cherry-picking" by the unscrupulous to distort the data set to support a particular viewpoint, SANS 241-2 suggests very specific guidelines for compacting the complete data set to five annual water quality indices to give a simpler, overall measure of the different types of risks:

- Microbiological compliance;
- Acute health chemical compliance;
- Chronic health chemical compliance;
- Operational compliance; and
- Aesthetic compliance

Each index is derived in two steps, namely, quantification followed by assessment. For quantification, ALL the samples of ALL the sampling points of ALL the determinands that have a bearing on a risk category are checked for their individual compliance. The index for the risk category is then calculated with:

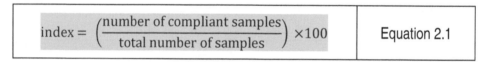

$$\text{index} = \left(\frac{\text{number of compliant samples}}{\text{total number of samples}}\right) \times 100 \qquad \text{Equation 2.1}$$

In the second step, the calculated index is categorised with the help of Table 2.4 to get one of three possible outcomes, namely, excellent, good or unacceptable.

Table 2.4: Assessment of water quality indices [a]

Risk Category	Water Quality Index		Assessment
	< 100 000 people	> 100 000 people	
Microbiological	≥ 97	≥ 99	Excellent
	≥ 95	≥ 97	Good
	< 95	< 97	Unacceptable
Acute health chemical	≥ 95	≥ 97	Excellent
	≥ 95	≥ 97	Good
	< 95	< 97	Unacceptable
Chronic health chemical	≥ 95	≥ 97	Excellent
	≥ 93	≥ 95	Good
	< 93	< 95	Unacceptable
Operational	≥ 93	≥ 95	Excellent
	≥ 90	≥ 93	Good
	< 90	< 93	Unacceptable
Aesthetic	≥ 93	≥ 95	Excellent
	≥ 90	≥ 93	Good
	< 90	< 93	Unacceptable

[a] Taken from Table 4 of SANS 241-2 (2015)

Example 2.3

(This example is a part of a larger, more detailed example in Annex B of SANS 241-2 (2015), which shows sample calculations for all the indices.)

Calculate the operational compliance of the drinking water of a town of 26 400 derived from three boreholes. The relevant determinands are (with the total number of samples and number of compliant samples in brackets):

- pH (1179; 1179)
- Turbidity (1179; 1085)
- Aluminium (16; 16)
- Free chlorine residual (1179; 1141)
- Total coliforms (16; 16)
- Heterotrophic plate count (136; 136)
- Somatic coliphages (16; 16)

The totals of all the above come to 3721 samples in total, with 3 589 in compliance.

> The overall compliance is (3 589/3 721) x 100 = 96.45.
>
> This is more than 93 (see Table 2.2 for towns with inhabitants fewer than 100 000) and the operational water quality is therefore declared as "excellent".

2.8. WATER QUALITY INDICES FOR OTHER PURPOSES

Because of the great many determinands for water quality, coupled with multiple sampling points and repeat measurements, the sheer number of data points for research projects could be overwhelming – similar to the monitoring required by SANS 241. Some go as far to call it "data-rich but information-poor" and rather opt for something in an "easily expressible and understandable format". Researchers frequently resort to their own versions of a "water quality index" (WQI) to reduce the data points to a single number on a scale from 0 to 100, with 100 indicating the best possible quality. Although questionable from a narrow scientific viewpoint, a single value scaled between 0 and 100 is useful for ranking purposes and moreover, comprehensible to most non-technical stakeholders. The utility of a simple WQI index was demonstrated in a recent case study which took spot samples from a few locations in Mpumalanga and North-West provinces (Wanda et al., 2015). By measuring seven variables (pH, DO, BOD, temperature, *E. coli*, nitrates, and phosphates), they demonstrated that their own WQI could detect the beneficial effects of treatment and that untreated water from natural surface water is unfit for potable purposes.

Researchers, who are mostly interested in answers to specific problems rather than always getting a full water quality profile with many determinands, also revert to simplified indices to illuminate specific issues. An assessment of the water supply systems of 15 rural villages in Venda used four indices to assess the water quantity, water quality, continuity of supply and physical condition of the systems (Haarhoff et al., 2008). To make a rapid spot check of the water quality, only four parameters were used – pH, electrical conductivity, turbidity and total coliforms. They were taken as simple indicators of corrosivity, brackishness, clarity and microbiological safety of the water in each system. To calculate a single-value WQI, they were weighted in the ratios 0.1; 0.1; 0.1; and 0.7. At a glance, the indices showed clearly that the biggest problem with the systems was continuity of supply due to many and prolonged supply interruptions. This simple WQI further showed that the water quality was generally good and that the failures were owing to microbiological contamination.

2.9. PRACTICAL IMPLEMENTATION OF A MONITORING PROGRAMME

This chapter demonstrates that the design of a proper water quality monitoring programme, even for a small community, requires careful planning and detailed knowledge of the latest requirements of SANS 241. Once the monitoring demands are defined in terms of the number and positions of sampling points, and the sampling frequency of the different determinands, the practical and economic implications can be quantitatively addressed. Some of the questions that must be addressed are:

- Should analyses be performed in-house or outsourced? The cost of setting up one's own ability for measuring a determinand can now be compared to the external cost.

- Can sampling be streamlined to allow for multiple determinands to be measured from the same sampling bottle? Taking different sample preservation measures into account, the number, type and size of sample bottles can be substantially consolidated.

- Consider a sensible sample numbering system with pre-printed sample bottle tags to avoid confusion in the laboratory.

- Always allow for some replicate samples to detect sampling and analytical errors.

- Invest in a LIMS (laboratory information management system) and feed it immediately with the numbers as they become available. Analytical data is not only about recording the water quality history (that too) but meant to be a "live" management tool for the water supply authority.

REFERENCES

Food and Agriculture Organization. 2001. *Manual on the Application of the HACCP System in Mycotoxin Prevention and Control*. Rome: Food and Agriculture Organization of the United Nations. http://www.fao.org/docrep/005/y1390e/y1390e0a.htm.

Haarhoff, J., Rietveld, L.C. and Jagals, P. 2008. "Rapid Technical Assessment of Troubleshooting of Rural Water Supply Systems." Proceedings of the 10th Annual Water Distribution Systems Analysis WDSA 2008. Van Zyl, J.E., Ilemobade, A.A., and Jacobs, H.E. Kruger National Park, 17–20 August 2008.

South African Bureau of Standards. 2015. *South African National Standard (SANS) 241-1. Drinking Water Part 1: Microbiological, Physical, Aesthetic and Chemical Determinands*. Pretoria: South African Bureau of Standards.

South African Bureau of Standards. 2015. *South African National Standard (SANS) 241-2. Drinking Water Part 2: Application of SANS 241-1*. Pretoria: South African Bureau of Standards.

Wanda, Elijah M.M., Mamba, Bhekie B. and Msagati, Titus A.M. 2015. "Determination of the Water Quality Index Ratings of Water in Mpumalanga and North West Provinces, South Africa." *Physics and Chemistry of the Earth* 92: 70–78. http://dx.doi.org/10.1016/j.pce.2015.09.009.

PART II

DRINKING WATER TREATMENT

CHEMICALS

3.1. INTRODUCTION

Drinking water treatment relies exclusively on physical-chemical processes, mediated by the addition of treatment chemicals. The proper control and management of chemicals at the treatment plant, therefore, is one of the prerequisites of professional plant operation. Other than the reactions when treatment chemicals are added to water, numerous considerations come into play. This chapter discusses the active ingredients in treatment chemicals, different methods to express their concentrations, accounting for impurities, ensuring the quality and safety of commercial products, dosing and inventory control for the different types of products, their safe handling and proper storage.

In the final section of the chapter, special emphasis is placed on the practical determination of chemical dosage at treatment plant level. The key to troubleshooting of treatment problems is the ability to rapidly and accurately determine, on site, the dosage rates of the different chemicals. All plants differ in some respects, so chemists and engineers must often work from first principles and need to hone their ability to determine chemical dosage concentrations fluently and accurately.

3.2. SPECIFICATION OF WATER TREATMENT CHEMICALS

3.2.1. Concentration of active ingredients

Treatment chemicals have one or more active ingredients – the remainder of the chemical remains passively in the background. In some cases, the chemical is entirely made up of the active ingredient, such as chlorine gas which consists of Cl_2 only. In most cases, however, the active ingredient is only a part of the chemical supplied. Chlorine, for example, could alternatively be supplied as a liquid or in solid form, with the concentration of active Cl_2 different in every phase. Therefore, one must know the concentration of the active ingredient in the commercial product to be able to properly adjudicate its price or determine its appropriate dosage. Four categories of chemicals are distinguished when considering their active ingredients:

- The easiest case is when the chemical is supplied as the active ingredient only, such as chlorine gas already mentioned. Powdered activated carbon is another example of a product supplied in pure or almost pure form.

- In some cases, the active ingredient is chemically bound within a larger molecule. Aluminium sulphate, ferric chloride and ferric sulphate are examples of commercial coagulants where the active ingredients are Al^{3+} and Fe^{3+}. The SO_4^- and Cl^- ions in the products serve no purpose. For the ferric products, for example, the active ingredient Fe^{3+} makes up 34% of pure ferric chloride and 32% of pure ferric sulphate, by mass.

- For reasons of long-term stability or solubility, chemicals are sold in diluted form. Solid material such as calcium hypochlorite has some inert material mixed in, while dissolved material is diluted by the solvent. The level of dilution therefore adds a further factor to consider when performing dosage calculations.

- The last category of chemicals has a secret "formula" for its chemical composition to protect its commercial interests. Dosage tests and calculations are then performed with the product itself, without knowing its constituents or active ingredients.

A special note is warranted in the case of "lime" which is used at many treatment plants for pH control, as it often means different things to different people. Be aware that limestone (predominantly calcium carbonate) is the raw material mined, which is then heated to drive off carbon dioxide to leave quicklime or unslaked lime (calcium oxide). The quicklime is then reacted with water (an exothermic reaction) to form slaked lime (calcium hydroxide). The available lime in slaked lime is the percentage of calcium hydroxide on a mass basis, typically from 75% to 95% because of unreactive contaminants in the original limestone and subsequent incomplete conversions to calcium oxide and calcium hydroxide.

Example 3.1

How much commercial sodium hypochlorite and commercial calcium hypochlorite must be added to water to have the same effect as 1 kg of chlorine gas? The concentrations of commercial products vary; so, they must be checked for specific applications. For this example, assume that commercial sodium hypochlorite contains 15% NaOCl (molecular mass 74 g/mol); commercial calcium hypochlorite contains 65% $Ca(OCl)_2$ (molecular mass 143 g/mol); and commercial chlorine gas contains 100% Cl_2 (molecular mass 71 g/mol).

In each kg (1000 g) of product, there is 150 g of NaOCl and 650 g of $Ca(OCl)_2$.

A mass of 150 g of NaOCl is equivalent to 150 x (71/74)/2 = 72 g of chlorine gas.

A mass of 650 g of $Ca(OCl)_2$ is equivalent to 650 x (71/143) = 323 g of chlorine gas.

To get the same effect as 1.0 kg of chlorine gas, we would need to add (1 000/142) = 13.9 kg of commercial sodium hypochlorite or (1000/323) = 3.1 kg of commercial calcium hypochlorite.

For proper scientific control and management of chemical dosing, it is essential to establish and maintain a consistent practice for the expression of dosage concentrations throughout the organisation. Misunderstanding often stems from two causes:

- Chemists are trained to express concentrations in mass or moles per litre, while operators prefer units corresponding to operational practice and instrument calibration at ground level. Units such as mL/minute, kg/h or even bags/shift or drums/day are encountered.

- Consistent units do not completely eliminate confusion as to what is really measured. From a chemical perspective, using aluminium sulphate as an example, we only really care about the active Al^{3+} concentration. But the concentration can also be expressed in terms of $Al_2(SO_4)_3$.

Furthermore, solid aluminium sulphate kibbles have 14 bound water molecules for each molecule of aluminium sulphate. Operators, weighing this material on site, would obviously prefer to work in terms of the $Al_2(SO_4)_3.14H_2O$ product. Should aluminium sulphate be used in liquid form, operators will likely prefer to work in terms of the mass or volume of solution added to the water. There are therefore four ways in which to express the dosage concentration of aluminium sulphate. The reported dosage concentration should always be fully disclosed as "mg Al/L" or "mg $Al_2(SO_4)_3.14H_2O$ /L", etc.

Example 3.2

Calculate the equivalent of 1 kg of Al (molecular mass 54 g/mol) for $Al_2(SO_4)_3$ (molecular mass 342 g/mol); $Al_2(SO_4)_3.14H_2O$ (molecular mass 594 g/mol); and a 50% solution of $Al_2(SO_4)_3.14H_2O$.

$Al_2(SO_4)_3$: (342 / 54) = 6.3 kg

$Al_2(SO_4)_3.14H_2O$: (594 / 54) = 11.0 kg

Liquid $Al_2(SO_4)_3.14H_2O$: (11.0 / 0.5) = 22.0 kg

3.2.2. Control of contaminants

All treatment chemicals, to a greater or lesser extent, are contaminated during mining, beneficiation or manufacture. The contaminants serve no useful purpose and leave less of the product to do useful work. Powdered lime with a 4% silica content, for example, only leaves 96% of the purchased product as useful $Ca(OH)_2$.

More importantly, detailed information on the impurities is required to determine whether their addition to water could compromise the treated water quality. A chemical with a high concentration of a particular contaminant could meaningfully elevate the concentration in the treated water. The concept of Recommended Maximum Impurity Content (RMIC) is used for quantitative guidance:

$$RMIC = \frac{NS}{MD \times SF} \times 10^6$$	Equation 3.1
RMIC = recommended maximum impurity content (mg/kg) NS = national water quality standard (mg/L) MD = maximum anticipated dosage (mg/L) SF = safety factor (-)	

A safety factor of 10 is judged to be reasonable, thereby limiting the contaminant to no more than 10% of the allowable national standard (World Health Organisation 2001).

> **Example 3.3**
>
> Manganese is a common contaminant in commercial lime, with levels in South African products reaching 6 000 mg/kg (Trollip et al., 2013). The SANS 241-1 standard allows 0.05 mg/L of manganese in drinking water. Does the manganese contamination pose a potential water quality problem?
>
> If the lime dosage is 10 mg/L, the RMIC = 500 mg/kg if the recommended safety factor of 10 is used.
>
> This is much higher than what was found by Trollip et al. (2013). Users of these lime products are therefore cautioned to ensure that the manganese is effectively removed during subsequent treatment.

3.2.3. Quality control of chemicals

Numerous international technical standards had been developed for different treatment chemicals to ensure their quality and safety. Although there are no locally developed South African standards, there is an institutional arrangement to allow the enforcement of the European (EN) standards. The standards are available from the SABS. Who should have the primary responsibility for the enforcement of the standards – the manufacturers, or the users? This question was considered for South African conditions (John and Trollip 2009). They concluded that it would be better to place the responsibility on the manufacturers on the following grounds:

- There are fewer manufacturers than end users, which makes regulation simpler.
- End users often do not have the expertise and means to perform their own testing.
- Analysis for contaminants in the product is much simpler than trying to find them in much lower concentrations after dilution in the treated water.

Water suppliers are therefore strongly encouraged to use approved products with proof of certification. End users could, of course, do their own tests for independent quality control. For this purpose, a manual of test methods was compiled for use in South Africa and is available from the Water Research Commission (Freese et al., 2004).

3.2.4. Specification of chemicals

As mentioned, it is apparent that the professional procurement of chemicals for water treatment is a technical process that requires expert scientific evaluation. The Water Research Commission addressed this matter by publishing a research report providing generic documentation to be used for the specification and procurement of treatment chemicals (John and Trollip 2009). This is a valuable document with a total of 44 pages devoted to the preparation of the tender documents and a further 14 pages to the structuring of the final contract between user and supplier — illustrating the need for technically competent staff to select, specify and procure the right chemicals.

3.3. COMMON TREATMENT CHEMICALS

3.3.1. Chemicals used in South Africa

In a rare survey in 2009, a total of 46 different treatment chemicals were identified by manufacturers and users that might be in use at South African water treatment plants (John and Trollip 2009). These chemicals have been grouped by operational categories, namely, for pH adjustment (Table 3.1), disinfection (Table 3.2) and coagulation (Table 3.3). Two more categories were listed which are not addressed here, namely, emergency disinfection (normally applied in pipelines or reservoirs after the water has been treated) and fluoridation (accepted in principle for South Africa years ago, but no indication yet of when it may be implemented).

Table 3.1: Chemicals used for pH adjustment[a]

Chemical Name (Synonym)	Formula (MW)	Phase (CAS Registry)
Sodium carbonate (soda ash)	Na_2CO_3 (106.0 g/mol)	Powder (CAS 497-19-8)
Calcium oxide (lime, quicklime)	CaO (56.1 g/mol)	Powder (CAS 1305-78-8)
Calcium hydroxide (slaked lime)	$Ca(OH)_2$ (76.1 g/mol)	Powder (CAS 1305-62-0)
Sodium hydroxide (caustic soda)	NaOH (40.0 g/mol)	Solution (CAS 1310-73-2)
Carbon dioxide	CO_2 (44.0 g/mol)	Gas (CAS 124-38-9)
Sodium hydrogen carbonate (baking soda)	$NaHCO_3$ (84.0 g/mol)	Powder (CAS 144-55-8)

[a] Selected from John and Trollip (2009)

Table 3.2: Chemicals used for disinfection[a]

Chemical Name (Synonym)	Formula (MW)	Phase (CAS Registry)
Sodium hypochlorite (liquid bleach)	NaOCl (51.0 g/mol)	Solution (CAS 7681-52-9)
Chlorine	Cl_2 (70.9 g/mol)	Gas (CAS 7782-50-5)
Anhydrous ammonia (ammonia gas)	NH_3 (17.0 g/mol)	Gas (CAS 7664-41-7)
Hydrogen peroxide	H_2O_2 (34.0 g/mol)	Solution (CAS 7722-84-1)
Potassium permanganate	$KMnO_4$ (158.0 g/mol)	Granular (CAS 7722-64-7)
Chlorine dioxide	ClO_2 (67.5 g/mol)	Solution (CAS 10049-04-4)
Ozone	O_3 (48.0 g/mol)	Gas (CAS 10028-15-6)

[a] Selected from John and Trollip (2009)

Table 3.3: Chemicals used for coagulation[a]

Chemical Name (Synonym)	Formula (MW)	Phase (CAS Registry)
Aluminium sulphate (alum)	$Al_2(SO_4)_3.nH_2O$	Kibbles (CAS 10043-01-3)
Sodium silicate (activated silica)	Na_2SiO_3 (122.0 g/mol)	Powder (CAS 1344-09-8)
Sodium aluminate	$Na_2Al_2O_4$ (163.9 g/mol)	Solid (CAS 1302-42-7)
Ferric sulphate	$Fe_2(SO_4)_3.nH_2O$	Solid (CAS 10028-22-5)
Ferric chloride	$FeCl_3.nH_2O$	Solution (CAS 7705-08-0)
Polyaluminium chloride	$Al_2(OH)_xCl_y.nH_2O$	Solution (CAS 1327-41-9)
Diallyldimethylammonium chloride	$(C_8H_{16}NCl)_n$ (161.7n g/mol)	Solution (CAS 26062-79-3)
Epichlorohydrin (dimethylamine)	$[C_5H_{12}OCl]_n$ (123.6n g/mol)	Solution (CAS 25988-97-0)
Polyacrylamides	$(C_3H_5NO)_n$ (71.1n g/mol)	Solution (CAS 9003-05-8)

[a] Selected from John and Trollip (2009)

3.3.2. CAS registration numbers

Tables 3.1 to 3.3 include the CAS registration number of each compound in its pure form. It refers to a number assigned to the compound by the Chemical Abstracts Service (CAS), a division of the American Chemical Society. CAS is the world authority for providing the global scientific community with access to the most current chemical information. The systematic collection of information on chemical compounds had a humble beginning in 1907 with a journal *Chemical Abstracts*. In 1965, a new era dawned with the introduction of the CAS Chemical Registry System. Since 2007, a web version of the CAS database is available. The unique CAS registry number identifies each chemical substance without the ambiguity of chemical nomenclature. The number of chemicals registered is staggering. At the start in 1907, less than 12 000 chemicals were covered. In 2009, 50 million were registered and new ones are now added at an average rate of 15 000/day. For getting detailed, authoritative and up-to-date information, www.cas.org must be consulted.

A CAS registry number has no inherent chemical significance and seems to be assigned in the order of being registered. The final digit is a check digit based on the digits before. For numerophiles:

- Take the second-last digit and multiply by 1
- Take the third-last digit and multiply by 2
- Take the fourth-last digit and multiply by 3
- ….. until you run out of digits
- Add all the calculated products
- The check digit is the sum of products MOD 10, for example 43 MOD 10 = 3; 370 MOD 10 = 0, 158 MOD 10 = 8.

The check digit for soda ash with CAS number 497-19-8 (first line in Table 2.1) is $(9x1 + 1x2 + 7x3 + 9x4 + 4x5)$ MOD $10 = (9 + 2 + 21 + 36 + 20)$ MOD $10 = 88$ MOD $10 = 8$.

The CAS database only considers chemical compounds in their pure form. Industrial chemicals, as pointed out before, are contaminated or diluted to some extent. The user of the CAS database must consider all the other compounds in the product where they may be relevant.

3.3.3. Chemical properties

Which physical and chemical properties are the most relevant to practical water treatment? The question is approached from different perspectives.

Dosage determination requires a relationship between what is measured in the laboratory and what is added to the water at the treatment plant. If the laboratory testing is done with the same product that is added on site, then it is a simple one-on-one relationship. For example, if 0.1 mL of a product is added to 1L of water, the same volumetric ratio must be maintained on site. However, laboratory testing is often done with different chemicals than used on site, for example:

- Tests with chlorine in the laboratory are done with bleach (precise dosing of gas in a laboratory is difficult), while chlorine gas is used on site.

- Tests with lime in the laboratory is usually done with a liquid lime solution (adding powder in small quantities to laboratory jars is tricky), while powdered lime is used on site.

In these cases, a chemical conversion must be made, which requires the identification of the active ingredient being tested, the mass concentration of the active ingredient in both products, and the molecular weight of both products to establish the molar equivalency between laboratory and treatment plant.

Example 3.4

A shock dose of HTH (active ingredient $Ca(OCl)_2$ with molecular mass 143.0g/mol) must be added to disinfect a small reservoir of 4 000 m^3 after a leak is repaired. Laboratory tests with liquid bleach (12% NaOCl solution with density of 1 050kg/m^3) indicate that 5 mL of bleach per litre of water will be required for proper disinfection.

Five mL of bleach = 5 x 1.05 = 5.25 mg, which contains 5.25 x 0.12 = 0.63 mg of NaOCl

0.63 mg of NaOCl = 0.63 / 51.0 = 0.0124 mmol

One mole of $Ca(OCl)_2$ is equivalent to 2 moles of NaOCl, therefore 0.0124 / 2 = 0.0062 mmol of $Ca(OCl)_2$ is required per litre of water to be treated.

Dosage control at the treatment plant often requires mass/volume conversion. The dosage stipulated may be in terms of a volumetric concentration, while the plant operators may be accustomed to working with mass concentration, or the other way around. The mass/volume conversion requires the specific gravity and the bulk density of the product used. Bear in mind that the specific gravity refers narrowly to the solid phase of the product, while the bulk density refers to the overall product. Grains of silica sand, for example, have specific gravity of about 2600 kg/m^3, but the density of a bag of filter sand (which includes the interstitial voids) is roughly 1600 kg/m^3.

Example 3.5

Continuing from Example 3.3, the problem is to add 0.0062 mmol of $Ca(OCl)_2$ per litre of water in the tank. Commercial calcium hypochlorite, called just "product" in this example, is a granular product with 70% $Ca(OCl)_2$ and bulk density of 800kg/m^3.

Molar concentration of $Ca(OCl)_2$ = 0.0062 mmol/L

Mass concentration of $Ca(OCl)_2$ = 0.0062 x 143.0 = 0.887 mg/L

0.887 mg/L $Ca(OCl)_2$ = 0.887 / 0.700 = 1.27 mg/L of product

A reservoir of 4000 m^3 requires a mass of 4 000 x 1.27 = 5 066 g of product

A mass of 5 066 g of product occupies a volume of 5 066/0.8 = 6 888 mL.

We therefore need 6 888/250 = 25.3 cups of product, if 1 cup = 250 mL

When solutions are prepared, the solubility product of the chemical is required to ensure that the solution strength is specified such that the chemical is readily and fully dissolved. The extent and nature of the impurities in the treatment chemical must be known to prevent its possible contamination of the treated water, a matter discussed in Section 3.2. The safe handling and storage of chemicals is discussed in Section 3.5.

3.3.4. Proprietary treatment chemicals

Some of the treatment chemicals are branded as specific commercial products, without disclosing their exact chemical ingredients. All laboratory testing and dosage on site can only be done in terms of the product as sold. Users also do not have recourse to detailed information normally available from the CAS Registry Number. In these cases, users have no other option than to rely on the chemical properties provided by the suppliers. It is suggested that all the required properties are identified in advance and requested.

3.3.5. Material Safety Data Sheets (MSDS)

The MSDS is an important way of communicating information about chemicals in the workplace. Although this seems an obvious need in a world where chemicals are widely traded and used mostly by non-chemists, it was only in the late 1960s that specific MSDS requirements were regulated for the first time, limited to the maritime industry. In 1983 the requirements were broadened to include the manufacturing industry and again in 1987 to include all economic sectors. The leading international agency for setting these standards is the Occupational Safety and Health Administration (OSHA), an agency reporting to the United States Department of Labor. We now have an international standard ISO 1-1014 that provides a recommended MSDS format.

The MSDS must be prepared by every manufacturer or importer of a chemical and be passed on to those who buy the chemical. There are therefore many different MSDS versions for the same chemical. The internet is a quick, convenient place to get the MSDS on all the treatment chemicals.

In South Africa, treatment chemicals are regulated by the Hazardous Chemical Substances Regulations of 1995 (amended in 2003), which require that employers take responsibility to train and inform workers about the chemical hazards they face in the workplace. In terms of these regulations, workers have the right to have access to the MSDS's held by the employer. Every MSDS should provide the information in Table 3.4.

Table 3.4: Mandatory information for the Material Safety Data Sheet

Chemical product and company identification	Physical and chemical properties
Composition, information on ingredients	Stability and reactivity
Hazards identification	Toxicological information
First aid measures	Ecological information
Fire-fighting measures	Disposal considerations
Accidental release measures	Transport information
Handling and storage	Regulatory information
Exposure controls, personal protection	Other information

3.4. TECHNOLOGIES FOR ADDING CHEMICALS TO WATER

3.4.1. Adding gases to water

Direct gas injection into water is an option used mostly for adding atmospheric air. The air may either be pumped through a diffuser at the bottom of a reactor (which requires some pumping energy), or the water can cascade down some engineered structure to entrain and dissolve air (which consumes some hydraulic head). If the gas is expensive, one would naturally prefer to dissolve all the gas in the water to avoid its loss to the atmosphere when the bubbles break the surface prematurely. This must be achieved by either producing very small gas bubbles with high specific surface, or by providing deep contact reactors to allow enough time for the rising bubbles to dissolve completely. Carbon dioxide, a highly soluble gas, dissolves quickly and can be added to water in this way in relatively shallow reactors. Ozone, less soluble, normally requires much deeper reactors, as much as 6m.

Direct air injection runs the risk of allowing undissolved gas to escape into the atmosphere. When atmospheric air is added, it is compressed on site and injected directly into the water. Other gases are usually bought in liquefied form in pressurised cylinders. In both cases, the gas flow must be precisely controlled and measured to maintain the required dosing concentration. For toxic gases like ozone, or potentially explosive gases like chlorine dioxide, the reactors must be enclosed, the off-gas must be monitored, collected and destroyed.

For highly toxic gases such as chlorine, direct gas injection cannot be used because of the unacceptably high safety hazard posed by undissolved chlorine gas coming off the water surface. The alternative is to dissolve the gas completely in a small volume of water (in other words, to make a strong solution of the gas) and to then add the solution to the bulk of the water being treated. The best, very common example of this approach is offered by a conventional gas chlorination system. A small side-stream of water is taken from the main flow and pumped at high velocity through an ejector. The ejector takes the form of a venturi-shaped obstruction in a pipe. Furthermore, the water is accelerated through the throat of the venturi to reach a pressure below atmospheric, utilising the Bernoulli principle. A second pipe is connected from the gas cylinder to the throat of the venturi, and the gas is therefore "sucked out" of the cylinder, to be completely dissolved in the highly turbulent side-stream. The side-stream of chlorine solution is then added and mixed into the main water flow. (For safety reasons, containers for toxic gases have regulating valves on their outlets which will not allow any gas to escape even if opened to the atmosphere – unlike liquid petroleum gas (LPG) cylinders used for cooking, for example. The gas will only flow from the cylinder if the pressure outside is reduced below atmospheric pressure.)

3.4.2. Adding liquids to water

Commercial liquid chemicals are highly concentrated to limit transportation costs. Treatment plants have the choice to use these chemicals directly in concentrated form, or to dilute the chemicals before dosing it to the water. Both options have some advantages and disadvantages to be considered:

- It is more difficult to achieve precise dosing control if very small volumes of liquid are delivered to the water. At small treatment plants, the use of undiluted chemicals literally translates to a drop every few seconds in some cases, encouraging the introduction of a dilution step in the interest of better precision.

- Dilution should not be carried too far, as the activity of some chemicals (notably polymers) decreases more rapidly if diluted beyond a certain point.

- Diluting the chemicals first adds to the possible margin of error, as it requires additional calculation, measuring exact quantities of chemical and dilution water, and controlled mixing to ensure complete dispersion.

Whether liquid chemicals are diluted or not, they are usually pumped with special chemical dosing pumps made of materials that are compatible with strong and often aggressive treatment chemicals. Moreover, chemical dosing pumps usually employ reciprocating pistons to deliver the chemical in pulses. It is therefore usual to always use these pumps together with a pulsation dampener to even out the flow of chemical into the water. Without dampening, when viewing the injection point at micro-scale, the main body of water will be either under- or overdosed – not conducive to effective coagulation.

3.4.3. Adding solid chemicals to water

Chemicals in particulate form are the most difficult to add to water. The product must first be added (by hand or with special handling equipment) to a holding tank to dissolve or to form into slurry. The slurry or solution is then added at the desired rate and mixed with the main water flow.

3.5. HANDLING AND STORAGE OF CHEMICALS

Safety is a key concern for handling and storing chemicals at treatment plants. Good management practice must address the following concerns:

- Is the container material chemically inert to the chemical stored? Ferric chloride, for example, is a highly corrosive liquid with pH lower than 2, which requires the storage tank to be rubber-lined with good quality control. Treatment chemicals are likely to be changed during the lifetime of a treatment plant, forcing the storage of "new" chemicals in "old" tanks – their compatibility should be checked first.

- Is the container structurally suitable for holding the chemical? Suppliers will usually supply chemicals in appropriate containers, but care should be taken when smaller quantities are repackaged by third parties for supply to small treatment facilities.

- Are precautions taken against chemical spills? A proper storage facility should be surrounded by bunt walls, designed to contain a complete spill of a storage tank within the bunt wall enclosure without contamination of the surrounding area or water body.

- Are proper monitoring, alarms and ventilation provided? Chlorine dioxide, above a certain critical air concentration, could combust spontaneously. The storage area for chlorine cylinders must be well ventilated with an audible detector to ensure the safety of operators.

- Are appropriate facilities provided for destruction of unwanted off-gas? Ozone gas, for example, must be collected and destroyed from ozone contact reactors.

- Are personnel provided with proper protective clothing and equipment? Besides appropriate items like boots, gloves, safety glasses and uniform (the PPE - personal protective equipment), emergency showers and eyewashes are required in the chemical storage area.

- Are personnel provided with adequate training and information? All chemicals have safety sheets to prescribe reasonable precautions when the chemical is handled or spilled. These sheets should be prominently displayed.

3.6. DETERMINATION OF DOSAGE CONCENTRATION ON SITE

3.6.1. Determination of the flow rate

The flow rate at a water treatment plant is best determined by a flow meter. Some meters provide an instantaneous rate of flow (usually in L/s), but all have totalisers which provide cumulative totals of the water volume passed through the meters. It is easiest to use a stopwatch to determine the time it takes for a fixed volume of water to pass the meter. Select the volume such that it takes at least three minutes to pass the water volume – this will limit the timing error to less than 1%.

For determination of the coagulant dosage, the raw water flow rate should be measured. Final chlorination takes place at the end of treatment, when some water has already been lost owing to treatment losses – about 8% to 10% of the raw water flow rate. Make sure that the flow meter used is appropriate for what you want to calculate.

In the absence of a flow meter, other options are available, such as measuring the rise rate in a reservoir of known area, or even measuring the overflow depths at all the filter overflow weirs and calculating the flow rate through each filter, then totalising. These approximations require more engineering judgment and not covered here.

In all cases, express the flow rate, in whichever way measured, in m^3/h to three significant digits.

3.6.2. Determination of the dosage rate

The easiest case is where a chemical of 100% purity is added to water, for example when chlorine gas is measured with a rotameter and added through an ejector to the water. The dosage rate is then read directly off the rotameter. This also applies to cases where the chemical composition of the chemical is unknown, typically for organic polymers. The dosage rate is expressed in terms of the mass of product as purchased. These polymers are in liquid form with a density normally close to 1000 kg/m^3.

The purity and concentration of the product must be taken into consideration, which is available from the supplier. The dosage rate of liquid ferric chloride (typically 43% concentration) must be multiplied by 0.43 to get the dosage of actual $FeCl_3$. Some research papers report the ferric chloride dosage in milligrams of Fe^{+3} while others go further and report it in moles. The latter two expressions are rarely, if ever used at treatment plants.

When intermediate solutions are made in holding tanks, dilution must be accounted for. The matter is further complicated when dissolving solid lumps of aluminium sulphate or ferric chloride. The lumps, even when dry, have 14 or 6 molecules of bound H_2O respectively. The dry mass of the lumps now must be converted from $Al_2(SO_4)_3.14H_2O$ to $Al_2(SO_4)_3$ or even to Al^{+3} if required ($FeSO_4.6H_2O$ to $FeSO_4$ or Fe^{+3} in the case of ferric chloride) before the dilution in the tank is factored in.

In all cases, express the dosage rate, in whichever way measured, in g/h to three significant digits.

3.6.3. Determination of the dosage concentration

The last step is to calculate the concentration of each chemical, simply by dividing the dosage rate by the applicable flow rate.

In all cases, express the concentration in mg/L to three significant digits.

Example 3.6

During a visit of five hours to a treatment plant, the water flow rate was reasonably constant. For better accuracy, the flow rates were determined every hour. (Pumping rates, even when the same pumps run all the time, vary slightly owing to changing tank levels.) The raw water flow rate was checked by stopwatch and calculated as 147 m³/h. The final water flow meter recorded a constant flow rate of 137 m³/h. Three treatment chemicals were added at the time of the visit:

- A lime slurry was added at the inlet, prepared with a screw feeder feeding lime from a bulk storage silo. The mass of the silo with lime can be read at a meter below the silo. At the start of the visit, the reading was 618.3 kg and at the end of the visit, it was 591.9 kg. The purity of the lime is 91%.

- Aluminium sulphate is prepared in a holding tank by adding 100L of 50% solution to water to make up 1200 L of solution. Over the duration of the visit, the level in the holding tank dropped by 268 mm. The tank is 1.0 m wide and 1.6 m long. The aluminium sulphate is added at the raw water inlet. The density of a 50% solution may be assumed to be 1 600 kg/m³.

- Chlorine is added to the final water. The chlorine gas flow rate was kept constant at 0.35 kg/h throughout the visit

First calculate the dosage concentration of $Ca(OH)_2$.

The mass of lime used is $618.3 - 591.9 = 26.4$ kg $= 26\ 400$ g in 5 h $= 5\ 280$ g/h.

After accounting for impurities, the dosage rate of $Ca(OH)_2$ is 5 280 x 0.91 = 4 805 g/h.

The concentration of $Ca(OH)_2$ is therefore 4 809 /147 = 32.6 mg/L.

Second calculate the dosage concentration of $Al_2(SO_4)_3$.

The mass of $Al_2(SO_4)_3$ added to the solution tank = 0.100 x 1 600 x 0.50 = 80 000 g.

The concentration in the solution tank is 80 000/1.2 = 66 667 g/m³.

The volume of solution used is 1.0 x 1.6 x 0.268 = 0.429 m³ in 5 h = 0.0858 m³/h.

The mass of $Al_2(SO_4)_3$ used is therefore 66 667 x 0.0858 = 5 717 g/h.

The concentration of $Al_2(SO_4)_3$ added is therefore 5 717 / 147 = 38.8 mg/L.

Finally calculate the dosage concentration of Cl_2.

The dosage rate of chlorine gas is 0.35 kg/h = 350 g/h.

The concentration of Cl_2 added is therefore 350 / 137 = 2.55 mg/L.

3.7. REFERENCES

Freese, S.F., Trollip, D.L., and Nozaic, D.J. 2004. *Manual for Testing of Water and Wastewater Treatment Chemicals*. Report 1184/1/04. Pretoria: Water Research Commission.

John, W. and Trollip, D. 2009. *National Standards for Drinking Water Treatment Chemicals*. Report 1600/1/09. Pretoria: Water Research Commission.

Leopold, P. and Freese, S.D. 2009. *A Simple Guide to the Chemistry, Selection and Use of Chemicals for Water and Wastewater Treatment*. Report TT 405/09. Pretoria: Water Research Commission. http://wrcwebsite.azurewebsites.net/wp-content/uploads/mdocs/TT-405-09.pdf

Trollip, D.L., Hughes, J.C. and Titshall, L.W. 2013. "Sources of Manganese in the Residue from a Water Treatment Plant." *Water SA* 39 (2): 265269. http://dx.doi.org/10.4314/wsa.v39i2.10

World Health Organisation (1997) "The WHO Guidelines for Drinking-Water Quality." http://www.who.int/water_sanitation_health/dwq/S02.pdf

COAGULATION AND FLOCCULATION

4.1. INTRODUCTION

The two primary purposes of water treatment are to make the water aesthetically attractive and to render it safe for human consumption. To reach the goal of aesthetically pleasing water, the solid suspended particles (usually expressed as turbidity) must be removed. After removing turbidity, it may be necessary to have additional steps for the removal of colour, odour or taste, but turbidity removal comes first. This chapter starts by exploring the nature and properties of typical particles found in raw water to demonstrate the need to grow the small raw water particles to larger flocs for better removal. This leads onto a discussion of the fundamental mechanisms whereby small particles can be made to attach and grow into flocs, called coagulation and flocculation. The proper use of the jar test is discussed next, a simple yet effective method universally used for selecting the most cost-effective coagulant and the optimal concentration to be added. The final part deals with the time and intensity of mixing as important determinands for effective coagulation and flocculation.

4.2. THE NATURE OF SUSPENSIONS

The behaviour of a suspension is determined by the size, density and surface charge of the primary particles. Once these are known, appropriate removal processes can be selected, followed by the design of suitably sized treatment reactors.

4.2.1. Particle size

Figure 4.1 shows a comparison of the typical sizes of microbiological particles commonly encountered in water, along with the ranges where different filtration technologies are appropriate. Not shown are the inorganic, mostly clay particle particles which are smaller than 2 μm. The particle removal capacity of conventional rapid sand filters, which are the backbone of almost all the existing treatment facilities in South Africa, is depicted with the label "granular filtration" in the topmost bar of Figure 4.1. Sand filtration is limited to removing particles of about 1 μm and larger. But Figure 4.1 also shows bacteria and viruses

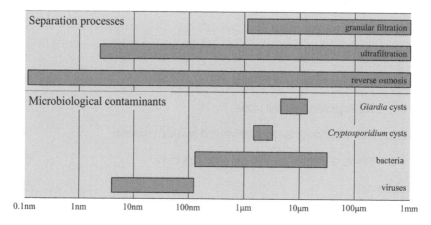

Figure 4.1   Removal range of particle separation processes

smaller than 1 μm which are not amenable to filtration. Coagulation and flocculation are essential to collect these small particles, along with larger particles, into flocs that are large enough to be removed by filtration.

Inspection by the human eye goes a long way to assess the degree of coagulation and flocculation in conventional treatment plants. A sample of 1 L, taken after 10 minutes of flocculation and observed through glass against a bright backlight provides useful qualitative information:

- If the water is somewhat murky or milky and no individual particles are visible, the coagulation and flocculation are not effective. The human eye can detect particles down to roughly 20 μm, often referred to as "pin floc".

- If large flocs are present (very easy to see) and they settle quickly to the bottom to leave the supernatant completely clear, it means coagulation and flocculation is effective and that the flocs are amenable to settling under gravity.

- If large flocs are present, but they settle slowly and incompletely, it means coagulation and settling is effective, but that the flocs do not settle readily – this is a situation where particle density must be considered more closely.

4.2.2. Particle concentration

Turbidity is the conventional and most common analytical method to quantify the suspended particle concentration in water. It is an indirect measure, as it only measures the attenuation of a light beam directed through the suspension. The strength of the light beam that reaches the opposite end of the suspension provides a turbidimetric measure of the particle concentration, while the amount of light that is reflected off the particles in a direction perpendicular to the light beam provides a nephelometric measure. Modern turbidimeters utilise both turbidimetry and nephelometry to report what is commonly just called turbidity, with units of NTU (from nephelometric turbidity units) or JTU (an older, almost extinct term derived from Jackson turbidity units where the light beam was produced with a "Jackson candle").

It stands to reason that turbidity, while it may provide a quick and easy measure of particle concentration, does not provide a sufficiently detailed tool for more in-depth analysis. Different types and sizes of particles may have different geometries and different reflective properties. A slightly better measure is to measure the suspended solids (SS), a gravimetric procedure not affected by the reflectivity of the particles, but still not taking size or shape into account. Based on experience at many South African water treatment plants, a rough correlation was observed of 1 mg SS/L ≈ 0.5 to 2.0 NTU.

Commercial particle counters provide a much more detailed characterisation of a suspension. These counters measure and count individual particles in a small sample of water flowing through its sensor. The user selects up to 32 size channels and the counter counts the number of particles in each range. Particle counters, although limited by their lower detection limit at around 1 μm, are powerful instruments for research purposes. They are, however, expensive and require careful operation. Ceronio et al (2002) provided a guide of how particle counting may be applied in the water industry. More importantly, the greater detail available from particle counters generates much more complex results. A single sample (for a counter with 32 channels) provides 32 numbers, which are difficult to interpret at a glance. For operational purposes, some simple data aggregation techniques are used, for example by reducing the 32 numbers to say "# particles < 10 μm" and "# particles > 10 μm" which are easier to comprehend and report. Researchers, on the other hand, not wanting to lose any of the detail, rather employ mathematical models with two or three model parameters to describe the particle numbers over all the sizes measured. Ceronio and Haarhoff (2005) compared alternative mathematical models proposed for this purpose.

4.2.3. Particle density

Filter sand has limited storage capacity for holding the flocs removed from the raw water. When the storage capacity is fully used, the filter must be backwashed to clear out the flocs from the openings among the sand grains. When the raw water has relatively few particles, such as clear water from mountain streams, the flocculated water can be applied directly to the sand filter without any processes between – the direct filtration option. In South Africa, such clear water is rarely available at municipal scale. Almost all raw water supplies in South Africa are subject to seasonal turbidity peaks, which disallow the use of direct filtration. If direct filtration would be used, the filters would block so rapidly that continuous production of filtered water would not be possible.

As a result, an extra phase separation process is required between flocculation and filtration. A choice must be made between sedimentation and dissolved air flotation, a choice guided by the density of the primary particles. Primary particles have densities varying from just below 1 000 kg/m³ (buoyant blue-green algae) to about 2 300 kg/m³ (clay particles). In the first case, the algal cells, even if they are well flocculated to produce flocs of 20 μm or more, will retain neutral buoyancy. In this case, settling will not work and dissolved air flotation should be used before filtration. In the case of clay particles, however, the larger, heavier flocs will settle readily making sedimentation the most effective process before filtration.

With eutrophic raw water, the raw water may carry predominantly clay particles during times of high river flow, but mostly algae during the low-flow periods. In such cases, there is a need for both dissolved air flotation and sedimentation. When the treatment plant has sedimentation facilities only, some clay particles are added to the raw water during times of algal growth to "weigh down" the flocs. The added clay, usually in the form of fine bentonite, is incorporated with the algal cells into the flocs, increasing the floc density to the point where they are successfully removed by sedimentation.

4.2.4. Particle charge

At micro-scale, coagulation/flocculation is a two-step process. The first step requires a transport mechanism to move the particles such that their trajectories intersect – in other words, forcing them closer together. This is a hydrodynamic issue dealt with by the hydraulic design of the coagulation/flocculation reactor. The second step requires an attachment mechanism to keep the particles together once they have touched. This is an issue dealt with by proper chemical conditioning of the primary particles.

The greatest hindrance to proper attachment is offered by the surface charge of the primary particles. If the particles had no surface charge, this concern would fall away. Unfortunately, almost all particles have some surface charge, mostly negative. Where such surface charge exists, it must be neutralised for successful flocculation to follow.

By applying an electrical field to a suspension and observing the direction in which the particles are moving with an optical microscope, it can be readily determined whether the surface charge is positive or negative. This principle of electrophoretic mobility underpins a method to measure the zeta potential of the particles in suspension, measured in millivolt (mV). Typical values for primary particles in water are about -10 mV to -20 mV. For successful coagulation, the zeta potential should be brought closer to 0 mV.

4.3. MECHANISMS FOR COAGULATION AND FLOCCULATION

4.3.1. Surface charge neutralisation

Almost all particles occurring naturally in water have negative surface charge, owing to the molecular structure of clay. By adding a cationic organic polymer, the polymer "coats" the surfaces of the particles and the cations of the polymer neutralise the negative charge on the particles. This is the coagulation step. The water is then slowly mixed to induce the particles to collide. Now, having no electrostatic repulsion, the particles "stick" together to form stable flocs. This is the flocculation step.

For surface charge neutralisation to work properly, a few requirements must be met. First, the initial mixing must be quite intense to disperse the polymer rapidly over the water volume, to give all particles equal exposure to the polymer. Second, the polymer dosage must be quite precise – too little will not neutralise the charges sufficiently; too much will lead to charge reversal on the particle surfaces, thereby restabilising the suspension.

Surface charge neutralisation works best at high particle concentrations when the water is turbid. Such dirty water has many particles and there are many contact opportunities when the water is mixed – therefore rapid flocculation. Conversely, if the water is very clear, there are only a few particles, far apart, leading to slower flocculation, often incomplete.

4.3.2. Sweep flocculation

Sweep flocculation is an alternative mechanism to surface charge neutralisation. It follows when the coagulant precipitates to form ferric or aluminium hydroxide flocs. Under favourable conditions, these chemical precipitates form large, "fluffy" flocs. As they move through the water, they "sweep up" all the small natural particles in their way. In this way, the particles are integrated into the flocs which are easily removed.

On the positive side, sweep flocculation does not require precise dosage as in the case of surface charge neutralisation. In addition, it is effective for both turbid and clear raw waters as the hydroxide flocs ensure adequate contact opportunities.

On the negative side, sweep flocculation produced more sludge than charge neutralisation, as both original particles and hydroxide flocs must eventually be removed.

4.4. THE JAR TEST PROCEDURE

How do we choose the "best" chemicals and their "right" dosage at a treatment plant? One cannot experiment directly on the plant itself, as all the water leaving the plant must comply with standards – once beyond the treatment plant, there is no opportunity to divert the water elsewhere before the customers are exposed. The time-tried jar test procedure is a simple, yet remarkably versatile laboratory tool for selecting the best coagulant or for selecting the most appropriate dosage. Although there are numerous different models on the market, jar testers all share the ability to test four to six jars at the same time to get a side-by-side comparison of the same chemical at different dosages. In a way, one can liken the jar test to a crude approximation of the coagulation, flocculation and sedimentation steps at a full-scale treatment plant. The speed of the mixing paddles can be controlled. Modern models have digital controls for programming the duration and intensity of consecutive mixing steps, as well as an illuminated base to allow better visual inspection of the suspension after mixing.

Jar testing has no standardised procedure, as users are free adopt their own "standard method" in order to allow comparisons with tests done at other times and to simulate full-scale mixing conditions where possible. For choosing the optimal dosing, the general pattern follows the same basic steps (assuming a six-beaker apparatus):

- Fill the beakers with the water to be tested. Leave the first (beaker #1) as control – no chemicals should be added.

- Start the stirring at high speed – typically 100 rpm.

- Decide on the range of dosage to be investigated. The lowest dosage goes to beaker #2 up to the highest dosage in beaker #6, with the "best" anticipated dosage in beaker #4.

- Add the chemical to be tested simultaneously to the different beakers with the stirrers turning. Start a stopwatch. The apparatus is now mimicking the coagulation step.

- After 1 to 2 minutes, slow the stirrers down to about 20 rpm. The apparatus is now mimicking the flocculation step.

- After another 5 to 15 minutes, stop the stirrers completely and pull them out of the beakers. The apparatus is now mimicking the sedimentation step.

- After another 15 to 30 minutes of settling, draw samples from about 30 mm below the water surface with a pipette and measure the desired response parameters (typically turbidity and pH as a minimum).

- Based on the results, narrow the range of dosages to be investigated and repeat.

Examples of jar testing are shown in short videos at https://www.youtube.com/watch?v=OWvM33L0-j4 and https://www.youtube.com/watch?v=kk9yquHvzXs.

4.5. OBJECTIVES OF JAR TESTING

Jar testing could be used for different purposes:

- For finding the optimal dosage of treatment chemicals. In some cases, the raw water quality remains constant for fairly long periods and jar testing at say weekly intervals would be sufficient. When there is a rapid change in raw water quality, for example after a flash flood, jar testing should follow immediately.

- For periodically checking whether alternative coagulants would not do better. Operators cannot quickly switch coagulants for reasons of large volumes of chemicals in storage or for annual chemical supply contracts, but they should periodically check whether other available coagulants would not do better. It is advisable to do comprehensive coagulant comparisons before supply contracts are awarded.

The process design for a new treatment plant requires numerous jar tests over a long period at different seasons to get a dependable water quality profile in terms of turbidity and other contaminants, stability, treatability and the 'best' chemicals for coagulation, disinfection, taste and odour control, etc.

4.5.1. Selecting the best dosage

How does one adjudicate the "best" dosage? The parameters below are all potentially important, but every situation will dictate its own yardstick(s):

- Turbidity after settling: this is the traditional yardstick and remains important.
- Turbidity after filtration: a simple test through a Whatman no. 1 filter paper gives a reasonable approximation of the expected filtrate quality after sand filtration.
- Removal of total or dissolved organic matter: this is a matter of growing importance, as we realise that organic matter is the precursor of a multitude of other water quality problems.
- Stability of the water: this is to ensure that the dosage strategy, mostly related to the use of metal-based coagulants and lime for pH correction, will render the water non-corrosive and non-precipitating once released into the water distribution network.
- Treatment residue production: residue management is becoming a major problem at water treatment plants, due to environmental concerns and the difficulty in obtaining discharge permits. It is a simple matter of measuring the mass of solids produced during jar testing to get a first estimate of anticipated residue production.
- Cost: the cost of treatment chemicals is a significant portion of total treatment costs and therefore of obvious concern. Jar testing allows the comparison of the "best" dosage of competing treatment chemicals on a cost basis.

4.5.2. Selecting the best coagulant

When comparing different coagulants, one would first select all the candidate coagulants. For each coagulant, a complete set of tests would be conducted to determine its "best" dosage, based on any one or more of the criteria listed in 4.5.1. After this, the "best" coagulant would be selected based on water quality, cost, and residue production. The product with the lowest dosage may not be the most cost effective owing to its high unit cost. A cheaper product requiring higher dosage may be more cost effective.

In practice, chemical suppliers send their representatives to treatment plants to do these tests with their own products, thereby doing some of the testing that normally would be performed by the treatment plant staff. The weakness of this practice is readily apparent, as competing products are not compared in a consistent, objective way – clients have no other option but to rely on the results submitted by the suppliers, each using their unique in-house testing procedures.

One of the problems encountered is the variability of the raw water quality. Some raw water supplies are quite consistent in quality and "snapshots" of water quality provided by jar tests provide a reliable basis for design and operation. However, in South Africa, we mostly deal with large fluctuations in water quality – dams go from overflowing to low levels regularly, rivers are steep and flow fast, etc. Raw water quality may change rapidly and widely and there is no guarantee that a few "snapshots" of jar tests will properly capture the raw water variability.

4.5.3. Enhanced coagulation

For many years, coagulants were only selected for their ability to remove turbidity. Towards the end of the 20th century, it was realised that coagulants, added in excess to what is required for the removal of turbidity, will also precipitate some of the dissolved organic carbon. This was quickly developed into a deliberate practice of enhanced coagulation, using metal-based coagulants at reduced pH, with proper guidance for its practical application. Not all raw waters are amenable to enhanced coagulation. Enhanced coagulation has a poor record in South Africa, which was eventually traced to the fact that the typical organic matter in our raw water sources are low-molecular weight compounds derived from return flows to our rivers, as opposed to the more humic, high-molecular weight compounds prevalent in the locations where enhanced coagulation was found to be effective.

4.6. FLOCCULATION MODELLING

4.6.1. Mixing intensity / velocity gradient

The intensity of mixing is a key concept for the agitation of liquids. It is commonly quantified as the root mean square velocity gradient or simply the G-value, normally denoted in seconds^{-1}. Some mixing systems derive the mixing energy from mechanical stirrers, while others get it from gravity as water flows from high to low. In both cases, the G-value can be calculated and compared:

$$G = \sqrt{\frac{P}{\mu \cdot V}}$$	Equation 4.1
G = velocity gradient (s^{-1}) P = energy input (Watt or kg.m^2.s^{-2}) μ = dynamic viscosity of water (kg.m^{-1}.s^{-1}) V = mixing volume (m^3)	

The energy input in the case of mechanical mixers relates directly to the power drawn by the electrical motor. For mixing that takes place when water drops by gravity into a mixing chamber, the energy input is provided by:

$P = \rho \cdot g \cdot Q \cdot H$	Equation 4.2

ρ = density of water (kg.m^{-3})

g = gravitational acceleration (m.s^{-2})

Q = water flow rate (m^3.s^{-1})

H = elevation difference in water level (m)

Example 4.1

Calculate the average G-value for a mixer drawing 2.1 kW in a mixing basin with volume 3.4 m^3 at a water temperature of 15°C. (The dynamic viscosity at 15°C is 0.00114 kg.m^{-1}.s^{-1})

Energy input = 2 100 W

Average G = SQRT (2 100 / (0.00114 x 3.4)) = 736 s^{-1}

Example 4.2

Calculate the a) energy input and b) average G-value for a flow of 430 L/s if the water level drops through 0.220 m while flowing through a hydraulic flocculator with volume 195 m^3 at a water temperature of 15°C.

Energy input = 1 000 x 9.81 x 0.430 x 0.220 = 928 W

G value = SQRT (928 / (0.00114 x 195)) = 64 s^{-1}

Coagulation normally requires a G-value of 500 to 1 000 s^{-1} for say 30 to 60 s to get rapid dispersal of the coagulant into the raw water; flocculation requires a much lower G-value of 30 to 50 s^{-1} of gentle mixing for about 8 to 12 minutes.

4.6.2. The Argaman-Kaufman model

The design of a flocculation system boils down to three fundamental decisions to be made by the process designer. Firstly, the type of system (backmix or plug flow) must be selected; secondly, the flocculation time must be decided upon; thirdly, the flocculation intensity (velocity gradient) must be chosen. Despite the crucial importance of the flocculation system to the rest of the treatment train, the designer is limited to relatively simple test procedures to guide him or her towards an optimal flocculation system. Moreover, these test procedures only provide "snapshots" of the flocculation behaviour at a particular combination of experimental variables — the flocculation behaviour at other experimental conditions (often) has to be crudely extrapolated or guessed.

A backmix reactor (often also called a CSTR or completely mixed stirred reactor) is normally a stirred tank where water enters somewhere and leaves somewhere else. Owing to tank geometry, the inlet and outlet are necessarily close to each other. It may therefore be that some water manage to leave the tank very soon after entering the reactor; other may take a much longer time. Conversely, a plug flow reactor has the inlet and outlet very far removed from each other (such as in a long pipe of channel) so that no water can suddenly make it from the inlet to the outlet. Plug flow reactors force all the water to have essentially the same hydraulic retention time in the reactor while backmix reactors are characterised by allowing a wide distribution of retention times. For flocculation, where time is of the essence to allow proper particle transport and aggregation, plug flow reactors are inherently more efficient.

The Argaman-Kaufman flocculation model can potentially be used as an extrapolation tool to generalise the flocculation behaviour of a particular coagulant / raw water combination from a small number of simple batch experiments. In other words, having only a few data points available, the designer can predict the flocculation behaviour at other flocculation times, at other velocity gradients, and even for other reactor types.

In essence, the model is based on two constants. The first is an *aggregation* constant K_a, which captures the ease at which particles will collide and attach. The second is a *breakup* constant K_b, which captures the ease at which flocs already formed will be broken apart. There are two versions of the model. The first, Equation 4.3, is applicable to backmix reactors:

$$\frac{n_0}{n_1} = \frac{1 + K_a . G . T}{1 + K_b . G^2 . T}$$

Equation 4.3

The second, Equation 4.4, applies to plug flow, or batch reactors with symbols the same as for Equation 4.3:

$$\frac{n_0}{n_1} = \left[\frac{K_b}{K_a} . G + \left(1 - \frac{K_b}{K_a} . G\right) . e^{-K_a . G . T} \right]^{-1}$$

Equation 4.4

Although the number of particles can be quantified with a particle counter, turbidity values have been used with success as a substitute parameter (Haarhoff and Joubert 1996).

4.6.3. Determination of the aggregation and breakup constants

Extensive tests have shown that good estimates of the constants can be obtained by performing nine jar tests. The mixing jars are plug flow reactors, where all the water has the same hydraulic retention time. The tests are performed at three velocity gradients (G = 30, 50 and 100/s) and three flocculation times (T = 4, 8 and 12 minutes). This leaves seven degrees of freedom (nine tests minus two constants). A few special precautions in conducting the tests are necessary to eventually find the constants by least-square regression.

4.6.4. Effects of flocculation time and mixing intensity

A typical set of Argaman-Kaufman model parameters, based on South African surface waters, would be (Haarhoff and Joubert, 1996):

*   Aggregation constant K_a = 9 x 10^{-5}
*   Breakup constant K_b = 6 x 10^{-8} s

Once the type of reactor is determined, the two remaining factors most critical for efficient flocculation are the mixing intensity and the flocculation time. Their effects on flocculation efficiency, along with typical design values for flocculation, are shown in Figures 4.4 and 4.5.

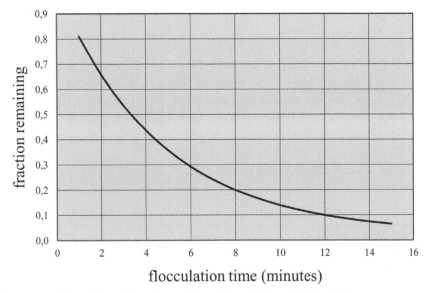

Figure 4.2   Effect of flocculation time on flocculation efficiency for G = 40 /s

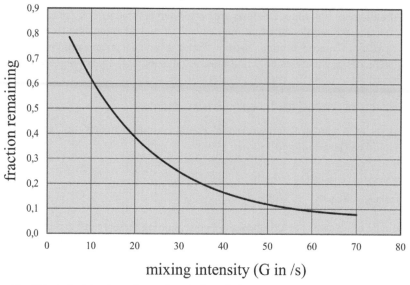

Figure 4.3   Effect of mixing intensity on flocculation efficiency for t = 9 minutes

---

Example 4.3

Use the Argaman/Kaufman model to determine the flocculation time if 80% of the turbidity must be removed at a mixing intensity of G = 35/s. Assume a completely mixed reactor and the typical constants provided in the text.

$n_1/n_0 = 0.20$

T = 1430 seconds = 23.8 minutes

Example 4.4

Repeat the previous problem but assume a plug flow reactor.

$n_1/n_0 = 0.20$

T = 540 seconds = 9.0 minutes

The results of this and the previous example indicate how much more effective plug flow reactors are in comparison with completely mixed reactors – the flocculation time is almost three times less.

4.7. REFERENCES

Ceronio, A.D., Haarhoff, J., and Pryor, M. 2002. *Standardisation of the Use of Particle Counting for Potable Water Treatment in South Africa.* Report TT 166/01. Pretoria: Water Research Commission. http://wrcwebsite.azurewebsites.net/wp-content/uploads/mdocs/TT-166-02.pdf

Ceronio, A.D., and Haarhoff, J. 2005. "An Improvement on the Power Law for the Description of Particle Size Distributions in Potable Water Treatment." *Water Research* 39 (2-3): 305313. https://doi.org/10.1016/j.watres.2004.09.023

Haarhoff, J., and Joubert, J.H.B. (1996) "Determination of Aggregation and Break-up Constants during Flocculation." Presented at Fourth International Conference of the IAWQ-IWSA Specialist Group on Particle Separation, 28-30 October 1996, Jerusalem, 916.

REFERENCES

STABILISATION

5.1. INTRODUCTION

"Stable" water, as discussed in this chapter, refers to water that will neither compromise the water reticulation system through excessive precipitation of solids on the inside of pipes nor damage the system through chemical attack or corrosion.

The stability and buffer capacity of natural waters are largely controlled by the carbonic system. The concepts of water stability are therefore mostly built around the idea that the water chemistry must be manipulated such that a thin film of calcium carbonate is maintained on the inside of pipes. The film, it is believed, must be maintained to protect the pipe material from corrosion. But at the same time, the rate of deposition must be controlled to prevent pipes from becoming partly or even completely blocked by chemical precipitates.

5.2. CARBONIC SPECIES IN WATER

The carbonic system consists of a connected series of chemical species:

- Dissolved carbon dioxide as CO_2. Carbon dioxide, in natural systems, dissolves from the atmosphere into water to establish equilibrium in accordance with Henry's law. Although the percentage of carbon dioxide in the atmosphere is small (0.03%); it is highly soluble in water.
- The dissolved CO_2 in the water maintains equilibrium with carbonic acid H_2CO_3.
- The H_2CO_3 partly dissociates into H^+ and bicarbonate HCO^3, depending on the pH.
- The HCO^3, in turn, partly dissociates to H^+ and carbonate CO_3^{2-}, depending on the pH.
- All the above species are in equilibrium with H^+ and OH^- ions.
- The relative concentrations of H^+ and OH ions determines the pH.

Figure 5.1 illustrates the main components of the carbonic system. The important point to realise is that the tendency to precipitate $CaCO_3$ depends directly on the pH. As the pH is raised, the tendency also increases. Our first aim is to determine the stability pH_s which is the pH where the water will neither dissolve nor deposit calcium carbonate. This is also the pH value that will be reached eventually if water is in contact with calcium carbonate long enough.

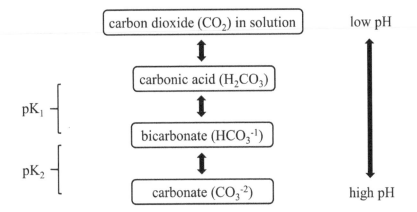

Figure 5.1   The carbonic system in water

5.3. Derivation of a practical expression for pH$_s$

At equilibrium, when water neither deposits $CaCO_3$ nor dissolves $CaCO_3$, the activities of the Ca^{2+} cations and CO_3^{2-} anions are related to its solubility product in Equation 5.1:

$$\{Ca^{2+}\}.\{CO_3^{2-}\} = K_s$$	Equation 5.1

The carbonate ions, being part of the carbonic weak acid system, maintain equilibrium with the bicarbonate ions according to Equation 5.2:

$$HCO_3^- \leftrightarrow H^+ + CO_3^{2-}$$	Equation 5.2

The equilibrium constant is provided by the second carbonic constant K_2 as shown in Equation 5.3:

$$K_2 = \frac{\{H^+\}.\{CO_3^{2-}\}}{\{HCO_3^-\}}$$	Equation 5.3

Rearrange Equation 5.3 to express the activity of the carbonate ions as:

$$\{CO_3^{2-}\} = \frac{K_2.\{HCO_3^-\}}{\{H^+\}}$$

Equation 5.4

Substitute Equation 5.4 into Equation 5.1:

$$\{Ca^{2+}\}.\frac{K_2.\{HCO_3^-\}}{\{H^+\}} = K_s$$

Equation 5.5

Rewrite Equation 5.5 as Equation 5.6:

$$\frac{1}{\{H^+\}} \cdot \frac{1}{K_s} \cdot \{HCO_3^-\} \cdot \{Ca^{2+}\} = \frac{1}{K_2}$$

Equation 5.6

In logarithmic format and with rearrangement, Equation 5.6 becomes:

$$\log\frac{1}{\{H^+\}} = \log\frac{1}{K_2} - \log\{HCO_3^-\} - \log\{Ca^{2+}\} - \log\frac{1}{K_s}$$

Equation 5.7

Rewrite Equation 5.7 in pX notation, where pX is the negative logarithm:

$$pH_s = pK_2 - pK_s - \log\{HCO_3^-\} - \log\{Ca^{2+}\}$$

Equation 5.8

Finally, it is necessary to account for the difference between molar concentration and activity. The molar concentration tells us how many ions are in solution while the activity roughly tells us what effect those ions have. When we titrate a solution, we measure the molar concentration directly, as each ion must be neutralised to get to the endpoint. Similarly, when we measure a compound with a gas chromatograph, we measure molar concentration as the detector senses every molecule as it leaves the chromatograph column. However, when we measure pH with a pH meter, we measure the effect that the hydrogen ions have on the membrane in the pH probe – therefore a direct measure of activity. In Equation 5.8, all concentrations are expressed as activities. The usual methods of measuring HCO_3^- and Ca^{2+} provide the molar concentrations, which must therefore be converted to activity. Equation 5.9 links the molar concentration to the activity with an activity coefficient:

activity {X} = activity coefficient f × molar concentration [X]	Equation 5.9

Equation 5.8 is therefore rendered as Equation 5.10:

$pH_s = pK_2 - pK_s - \log(f_m.[HCO_3^-]) - \log(f_d.[Ca^{2+}])$	Equation 5.10
f_m = monovalent activity coefficient (-) f_d = divalent activity coefficient (-)	

Further rearrangement provides Equation 5.11, useful for practical application:

$pH_s = pK_2 - pK_s - \log[HCO_3^-] - \log[Ca^{2+}] - \log f_m - \log f_d$	Equation 5.11

The stability pH_s predicted by Equation 5.11 is not something that can be measured. It is a calculated, theoretical value which simply gives the pH at which solid calcium carbonate will neither be dissolved nor precipitated.

5.4. ASSEMBLING THE CONSTANTS

The calculation of the pH_s requires six input values. Two of these values, $[Ca^{2+}]$ and $[HCO_3^-]$, are measured analytically. The other four are constants which are obtained from published correlations, presented below.

5.4.1. Correlation for pK_2

The second carbonic constant had been determined experimentally and modelled with the correlation shown in Equation 5.12 (Harned and Davis 1943):

$pK_2 = -6.4980 + 0.02379 \times t + \dfrac{2902.39}{t}$	Equation 5.12
t = water temperature (degrees K)	

Example 5.1

Calculate the values for pK_2 for water temperatures of 5°C and 20°C.

For 5°C = 278 K: pK_2 = -6.498 + 6.617 + 10.435 = 10.553
For 20°C = 293 K: pK_2 = -6.498 + 6.974 + 9.901 = 10.376

5.4.2. Correlation for pK_s

Calcium carbonate can precipitate in three forms – calcite, aragonite and vaterite. For drinking water purposes, only calcite is relevant. Its solubility product had been determined experimentally and the resulting correlation is shown as Equation 5.13 (Plummer and Busenberg, 1982):

$pK_s = 171.9065 + 0.077993 \times t - \dfrac{2839.319}{t} - 71.595 \times \log t$	Equation 5.13

Example 5.2

Calculate the values for pK_s for water temperatures of 5°C and 20°C.

For 5°C = 278 K: pK_s = 171.907 + 21.682 - 10.213 – 174.981 = 8.394
For 20°C = 293 K: pK_s = 171.907 + 22.852 - 9.691 – 176.615 = 8.454

5.4.3. Correlation for activity coefficients

Both the water temperature and the charge on the ions influence the activity coefficients. For water with low ionic strength (less than 0.5M which is usual for drinking water), the activity coefficients are estimated with the Davies correlation shown in Equation 5.14:

$\log f_i = -A.z^2.\left(\dfrac{\sqrt{\mu}}{1+\sqrt{\mu}} - 0.3.\mu\right)$	Equation 5.14
μ = ionic strength (mol/L) $z = 1$ for monovalent ions; $z = 2$ for divalent ions	

Equation 5.15 provides the constant A:

$A = 1.825\times10^6 . (78.3\times t)^{-1.5}$	Equation 5.15

In the rare event where the full ionic composition of the water is known, it can be computed from first principles as the sum of all the ions, shown as Equation 5.16:

$\mu = \dfrac{1}{2} . \sum C_n . z^2$	Equation 5.16

In practice, however, it is simpler and more common to measure the total dissolved solids (TDS) and to estimate μ with the correlation shown in Equation 5.17:

$\mu \cong 2.5\times10^{-5} . TDS$	Equation 5.17
TDS = suspended solids (mg/L)	

Example 5.3

Calculate the monovalent and divalent activity coefficients at 5°C and 20°C for TDS = 240 mg/L.

For TDS of 240 mg/L: $\mu = 0.00600$

For 5°C: A = 0.568

For 5°C and z = 1: log f_m = -0.040 thus f_m = 0.91

For 5°C and z = 2: log f_d = -0.037 thus f_m = 0.92

For 20°C: A = 0.525

For 20°C and z = 1: log f_m = -0.159 thus f_d = 0.69

For 29°C and z = 2: log f_d = -0.147 thus f_d = 0.71

5.4.4. Expression for log [HCO$_3^-$]

The bicarbonate ion is usually measured by acid titration to an endpoint of pH 8.3 by using phenolphthalein as indicator. This is called alkalinity and reported in mg/L as CaCO$_3$. For a refresher see http://nitttrc.ac.in/four%20quadrant/eel/quadrant%20-%201/exp7_pdf.pdf

The measured alkalinity must therefore be converted to a molar concentration by dividing it by the molar mass of $CaCO_3$ (100 g/mol), secondly divided by 1 000 to convert from mmol to mol, and finally multiplied by 2 to account for the difference in charge between CO_3^{2-} and HCO_3^-:

$$\log[HCO_3^-] = \log\left(\frac{2.\text{alkalinity}}{100 \times 1000}\right)$$	Equation 5.18

5.4.5. Expression for log [Ca²⁺]

Calcium in water can be measured in different ways. It is a test frequently performed because calcium is an important part of water hardness. Numerous test kits and testing strips had been developed and are readily available. The concentration is conventionally reported in mg/L as Ca and must be converted by dividing by the molar mass (40 g/mol) and 1000 to convert from mmol to mol:

$$\log[Ca^{2+}] = \log\left(\frac{\text{calcium}}{40 \times 1000}\right)$$	Equation 5.19

Example 5.4

Express an alkalinity measurement of 110 mg/L and calcium measurement of 42 mg/L in log molar concentration.

Alkalinity of 110 mg/L: log $[HCO_3]$ = -2.658

Calcium of 42 mg/L: log [Ca] = -2.979

Example 5.5

Assemble the answers of Examples 5.1 to 5.4 and calculate the stability pH_s at 5°C and 20°C.

For water at 5°C: pK_2 = 10.553; pK_s = 8.394; log $[HCO_3]$ = -2.658; log [Ca] = -2.979; log f_m = -0.040; log f_d = -0.159, therefore pH_s = 7.99

For water at 20°C: pK_2 = 10.376; pK_s = 8.454; log $[HCO_3]$ = -2.658; log [Ca] = -2.979; log f_m = -0.037; log f_d = -0.147, therefore pH_s = 7.74

5.5. CALCIUM CARBONATE PRECIPITATES

The pH_s, as calculated, is based on the solubility product for calcite, one of the three solid forms of calcium carbonate. At normal water temperature, it is the most dominant form, illustrated by Figure 5.2.

Figure 5.2   Electron micrograph of calcite crystals (Kimpton, 2016)

The second form, aragonite, forms at higher temperatures and is more likely to be found in hot water systems. It is shown in Figure 5.3.

Figure 5.3   Electron micrograph of aragonite crystals on sand grains (Buckman, 2020)

The third form is vaterite, a metastable form which is not likely to be found in water distribution systems.

5.6. OPERATIONAL CONTROL OF WATER STABILITY

If we could maintain water at exactly the pHs value, then it would theoretically neither attack nor deposit calcium carbonate. However, we wish to raise the pH slightly above the pH_s to deposit a thin protective layer calcite on the inner pipe surface. It was this thinking that led Professor Wilfred Langelier, a chemist at the Department of Civil Engineering at the University of Berkeley in California, to publish a landmark paper in 1936 where he proposed to keep the $(pH - pH_s)$ or Langelier Index as it became known, at about +0.2 to +0.3 units. In other words, if the pH_s is 8.1, then the water pH should be adjusted to about 8.3 to 8.4.

After being used for many years all over the world, it was eventually realised that the Langelier Index was a good measure of the *tendency* of calcite to precipitate, but not necessarily of the *amount* of calcite deposited. What is required in practice is a control on the amount of calcite deposited in the pipes. From this thinking, the Calcium Carbonate Precipitation Potential (CCPP) was developed, which predicts, in mg/L, how much calcite will be precipitated. A practical guideline for operational control of water stability is to keep the CCPP between 2 and 6 mg/L. To illustrate the difference, Table 5.1 shows five different water sources which all have the same Langelier Index of +0.3, but with widely different capacity to precipitate calcium carbonate, expressed as the CCPP.

Table 5.1: Illustration of the limitation of the Langelier Index[a]

pH	Alkalinity (mg/L)	Calcium (mg/L)	Langelier Index	CCPP (mg/L)
7.3	300	460	+ 0.3	35
7.9	188	190	+ 0.3	8
8.4	95	95	+ 0.3	2
9.0	65	19	+ 0.3	5
8.4	38	40	+ 0.3	2

[a]Taken from Loewenthal et al. (1986: Table 4.3)

A further limitation of the pH_s concept, as developed, is that it neglects the potential role of other ions. A good example is the presence of sulphates, abundantly present in acid mine drainage. In this case, the calcium ions will not remain as a static quantity in the background as assumed in the earlier derivation, but also maintain another equilibrium with calcium sulphate. To account for the presence of sulphates, the calculation becomes considerably more complex. By adding more other ions, the calculations become intractable to do by hand. Fortunately, modern chemical speciation software can factor in these complications.

The program called STASOFT was developed in South Africa in the late 1980s by researchers at the University of Cape Town, under the leadership of the late Prof RE Loewenthal, specifically with the drinking water industry in mind. The program has now reached version V (released in 2011) and available at http://www.wrc.org.za/knowledge-hub/e-tools/

5.7. WATER STABILITY AT THE TREATMENT PLANT

Surface water from dams and rivers are generally stable. Being in contact with sediments, sand and rocks for long periods, the water had plenty of opportunity to dissolve calcium carbonate if it is aggressive, or to deposit calcium carbonate if it is supersaturated. So why are we so concerned with water stability at the treatment plant?

5.7.1. Unstable raw water

Groundwater reaches its own equilibrium with surrounding rocks and sand in the aquifer, often at pressures and temperatures quite different than on the surface. Oxygen may be limited or absent, or other gases such as carbon dioxide might be present in high concentration. When the groundwater arrives at the surface and exposed to the atmosphere, it may be unstable and in need of pH correction.

5.7.2. Water with insufficient buffer capacity

Water must have a certain minimum buffer capacity before it can be safely distributed. There would be no point in carefully manipulating the pH during treatment if slightly different conditions in the distribution system could easily elevate or depress the pH. A minimum alkalinity of 30 to 35 mg/L as $CaCO_3$ is often used as a guideline. This is a problem prevalent in the Southern Cape along the coast from roughly Humansdorp to Cape Town.

5.7.3. Water softening

Water hardness is caused by divalent cations, predominantly Ca and Mg. Too much hardness causes domestic problems such as soap and shampoos that do not lather and scale formation in household devices. Industrially, it has severe economic consequences because of equipment damage and the need for frequent cleaning. The removal of hardness is commonly done by raising the pH to about pH 11, the point where $CaCO_3$ and $Mg(OH)_2$ are precipitated and removed by settling. Afterwards, the pH is depressed again until the required stability is reached.

5.7.4. Compensating for acidic treatment chemicals

Some treatment chemicals interact with the carbonic system. In other words, a stable raw water might not be stable after treatment chemicals are added. Most serious are the metal coagulants used for sweep flocculation:

$$FeCl_3 + 3HCO_3^- \rightarrow \underline{Fe(OH)_3} + 3CO_2 + 3Cl^-$$

$$Al_2(SO4)_3.14H_2O + 6HCO_3^- \rightarrow 2\underline{Al(OH)_3} + 12H_2O + 6CO_2 + 3SO_4^{-2}$$

The reactions show that 1 mole of $FeCl_3$ consumes 3 moles of HCO_3^-, while 1 mole of $Al_2(SO_4)_3$ consumes 6 moles. Table 5.2 presents the relationships in terms of the mass concentrations.

Table 5.2   Effect of acidic chemicals on alkalinity

Chemical	Alkalinity Consumed [a]
Ferric chloride	0.93
Ferric sulphate	0.53
Aluminium sulphate	0.51
Chlorine	1.41

[a] Expressed as mg $CaCO_3/L$ for every mg chemical/L

The acidifying effect of the chemicals must be counteracted by the addition of lime or other alkaline agents, shown in terms of the mass concentrations in Table 5.3.

Table 5.3   Effect of alkaline chemicals on alkalinity

Chemical	Alkalinity Added [a]
Sodium hydroxide	1.25
Lime	1.35
Soda ash	0.94
Sodium bicarbonate	0.60
Sodium hypochlorite	0.67

[a] Expressed as mg $CaCO_3/L$ for every mg chemical/L

Example 5.6

Assume that raw water is perfectly stable before treatment. If we add 22 mg/L of ferric chloride, how much lime must be added to return the water to the same alkalinity concentration?

22 mg/L ferric chloride will reduce the alkalinity by 0.93 x 22 = 20.5 mg/L

To restore the alkalinity, 20.5 / 1.35 = 15.2 mg/L of calcium hydroxide must be added

5.8.   CONVERSION OF UNITS

Although somewhat off the topic of this chapter, this section on the conversion of units is added at the end of the first five chapters, all dealing with chemicals and chemical treatment. It should be evident by now that the practicing chemist at a water treatment plant will be required to perform dosing calculations frequently, and that they should be done fluently and accurately.

Some units are intuitively easy to convert. For example, everyone can convert say inches to mm if we know the conversion factor of 25 mm ≈ 1 inch. However, when there are two sets of units involved, like say converting miles per hour to metres per second, it is slightly trickier, even if you know the conversion factors. With three or more sets of units, the problem quickly becomes intractable and a more systematic approach is required. Moreover, when you do these conversions, you also must leave an audit trail of what you did – you or someone else might want to check your calculations. Follow these steps, write them out and keep a copy.

Step 1
Write the units of the starting number clearly separated above and below the line.

Step 2
Write the units of the desired final number next to it.

Step 3
Check for parity between the sets of numbers. Where units are the same, there is no need to convert. Where they are not the same, write down the "path" of the conversion – to go from "marathon" to "metres", for example, we would follow the path marathon → kilometres → metres. Or to go from "day" to second" we could follow day → hour → minute → second. Or from "volume" to "weight" we could use volume (in cubic metre) → mass (in kilogram) → weight (in Newton).

Step 4
Write down the starting number with its units clearly separated above and below the line. Now convert the units that need conversion one by one by cancelling the unwanted units until you get the desired units. Leave the numeric parts in their original form. Example – convert 34 km/day to mm/s:

$$34 \, \frac{\text{km}}{\text{day}} \times \frac{1000 \text{ m}}{1 \text{ km}} \times \frac{1000 \text{ mm}}{1 \text{ m}} \times \frac{1 \text{ day}}{24 \text{ h}} \times \frac{1 \text{ h}}{3600 \text{ s}} = \left(\frac{34 \times 1000 \times 1000}{24 \times 3600} \right) \frac{\text{mm}}{\text{s}}$$

Step 5
Do the calculation last – it is mostly possible to simplify manually before you need your calculator.

$$34 \, \frac{\text{km}}{\text{day}} = 394 \, \frac{\text{mm}}{\text{s}}$$

A final suggestion. It is always useful to have the most basic conversion units in your memory. Know the tera-, giga-, mega-, kilo-, hecto-, deka-, X-, deci-, centi-, milli-, micro-, nano- sequence by heart. It saves time to know that 1 cubic metre of water has a mass of 1000 kg, that 1 litre of water has a mass of 1 kg, that there are 86400 seconds in a day, etc.

5.9. REFERENCES

Buckman, J. 2020. Electron micrograph of aragonite crystals. https://www.hw.ac.uk/schools/energy-geoscience-infrastructure-society/about/facilities/esem.htm

Harned, H.S., and Davis, R. 1943. "The Ionization Constant of Carbonic Acid in Water and the Solubility of Carbon Dioxide in Water and Aqueous Salt Solutions from 0 to 50°." *Journal of the American Chemical Society*, 65(10): 20302037. https://doi.org/10.1021/ja01250a059.

Kimpton, C. 2016. Electron micrograph of calcite crystals, entered in the 2016 competition of the Royal Photographic Society.

Loewenthal, R.E.; Wiechers, H.N.S. and Marais, G.v.R. 1986. *Softening and stabilisation of municipal waters*. Report TT 24/86. Pretoria: Water Research Commission.

Plummer, L.N., and Busenberg, E. 1982. "The Solubilities of Calcite, Aragonite and Vaterite in CO2-H2O Solutions Between 0 and 90°C, and an Evaluation of the Aqueous Model for the System CaCO3-CO2-H2O." *Geochimica et Cosmochimica Acta*, 46(6):10111040. https://doi.org/10.1016/0016-7037(82)90056-4.

Water Research Commission. STASOFT software for water stability calculations. http://www.wrc.org.za/knowledge-hub/e-tools/

DISINFECTION

6.1. INTRODUCTION

Drinking water treatment always includes some form of disinfection. Inevitably, some micro-organisms are not completely trapped or removed by conventional phase separation processes and must be inactivated by disinfection. Initially, the guidelines for disinfection were empirical, simply based on dosage ("add X mg/L of chlorine") or, somewhat more sophisticated, based on residuals ("add chlorine to maintain a residual of X mg/L of chlorine in the receiving reservoir"). During the 1990s, a systematic framework was developed for the rational selection of disinfectants and the determination of their minimum dosages to safeguard public health. At its core are the Log Removal and CT concepts. In most developed countries, this methodology has been incorporated in their national standards. But all chemists and engineers should heed these concepts, whether they are legislated or not, and apply them in the interest of public health.

6.2. THE CHICK-WATSON LAW

In 1908, two scientists Harriette Chick and Herbert Watson independently published papers which eventually led to what is known today as the Chick-Watson Law, which links the inactivation of micro-organisms with both the disinfectant dosage and contact time:

$$\ln \frac{N}{N_0} = K_{CW} \cdot C^n \cdot T$$	Equation 6.1
N = surviving number of organisms (-) N_0 = initial number of organisms (-) K_{CW} = Chick-Watson coefficient of specific lethality (L.mg^{-1}.min^{-1}) C = concentration of disinfectant (mg/L) n = coefficient of dilution, frequently assumed to be 1 (-) T = contact time (minutes)	

Example 6.1

An initial number of 17 000 organisms are reduced to four after exposure to 3 mg/L of disinfectant for 3 minutes. Assuming the same organisms, disinfectant and a coefficient of dilution of 1, use the Chick-Watson Law to calculate the concentration required to reduce an initial population of 24 000 to two after 15 minutes.

From the known performance of the disinfectant, calculate the Chick-Watson coefficient as

K_{CW} = -0.928 L/(mg.min). Use this coefficient to calculate the dosage requirement as C = 0.84 mg/L

The Chick-Watson model is expressed in terms of the survival ratio N/N_0. Instead of working with the number of surviving organisms, the water industry prefers to work with the removal ratio $(1 - N/N_0)$.

Example 6.2

What are the removal ratios for the two scenarios used in Example 6.1?

$N = 4$ and $N_0 = 17\,000$, therefore removal ratio $= 1 - 4/17000 = 0.999765$ or 99.9765 %

$N = 2$ and $N_0 = 24\,000$, therefore removal ratio $= 1 - 2/24000 = 0.999917$ or 99.9917 %

Another way to express the removal is to calculate the log removal with Equation 6.2:

$$\text{log removal} = -\log_{10}\frac{N}{N_0} = \log_{10} N_0 - \log_{10} N \qquad \text{Equation 6.2}$$

Example 6.3

What are the log removals for the two scenarios used in Examples 6.1 and 6.2?

Log removal $= \log 17000 - \log 4 = 3.63$

Log removal $= \log 24000 - \log 2 = 4.08$

There is a simple relationship between log removal and removal ratio:

- Removing 9 organisms from 10 equals a removal ratio = 90% or log removal = 1
- Removing 99 organisms from 100 equals a removal ratio = 99% or log removal = 2
- Removing 999 organisms from 1000 equals a removal ratio = 99.9% or log removal = 3
- Removing 9999 organisms from 10000 equals a removal ratio = 99.99% or log removal = 4

6.3 LOG REMOVAL REQUIRED BY DISINFECTION

Drinking water might harbour many pathogenic species, each with its own infectivity, virulence and resistance to disinfection. Individual monitoring of all potential organisms is clearly impractical. The problem is simplified by considering the contaminants in their conventional microbiological classes, namely viruses, bacteria and protozoa. For each group, one or more reference organisms are singled out for posing the largest threat. The reference organisms are those that have high pathogenicity, poor removal during treatment and long survival in the environment. The thinking is that if these reference organisms are inactivated, the risk posed by the hundreds of other species in the group should also be eliminated.

Once a reference organism is agreed on, the question arises whether it can be measured easily and whether it occurs in high enough concentrations to be statistically meaningful. If not, a surrogate or indicator organism is routinely measured in lieu of the reference organism, even if the surrogate organism is not pathogenic at all. Typical reference and surrogate organisms are shown in Table 6.1.

Table 6.1: Reference and surrogate organisms for three microbiological classes[a]

Hazard	Reference	Surrogate / Indicator
Protozoa	*Cryptosporidium parvum*	*Clostridium perfringens*
Viruses	Amalgam of rotavirus and adenovirus	Somatic / F-RNA coliphages
Bacteria	*Campylobacter*	*E. coli* / enterococci

[a] Table 4 from New South Wales (2015)

What should the log removal for different organisms be? Different authorities may have different requirements. The log removals legislated in the USA, for example:

- Viruses in surface water: 4 log removal AND *Giardia* 3 log removal
- Viruses in groundwater: 4 log removal

These values apply to the overall treatment train. The first part of the process train (flocculation, settling or dissolved air flotation, and filtration) removes a significant portion of the organisms in the raw water. These processes play a part in reducing the health risk and should get some credit for the overall removal of the organisms. Typical treatment credits are shown in Table 6.2.

Table 6.2: Log removal credits for physical treatment barriers[a]

Technology	*Cryptosporidium*	*Giardia*	Viruses
Settling / filtration	more than 2	2.5	2.0
Slow sand filtration	more than 2	2.0	2.0
Direct filtration	more than 2	2.0	1.0
Microfiltration	more than 2	more than 3	0
Ultrafiltration	more than 2	more than 3	0
Nanofiltration	more than 2	more than 3	3.0
Reverse osmosis	more than 2	more than 3	3.0

[a] Taken from Table 1 of USEPA Surface Water Treatment Rule Fact Sheet n.d.: 4

The log removal targeted for disinfection is obtained by subtracting the treatment credits from the overall log removal requirement.

Example 6.4

Determine the minimum log removal for viruses by disinfection at a) a direct filtration plant and b) a conventional filtration plant. Take the credits for treatment from Table 6.2.

Direct filtration: Log removal required = 4 (overall) – 1 (credit) = 3

Conventional filtration: Log removal required = 4 (overall) – 2 (credit) = 2

6.4. DETERMINATION OF THE CT PRODUCT REQUIRED

The basic premise of the CT concept is encapsulated in Equation 6.3:

log removal $\propto$ CT	Equation 6.3
C = disinfectant concentration (mg/L) T = contact time (h)	

Equation 6.3 is a simplification of the Chick-Watson Law after setting the coefficient of dilution n = 1. Regulatory authorities have published elaborate tables from which the required CT can be read. The required CT products are different for each disinfectant, which are discussed under separate sub-headings.

6.4.1. Free chlorine

The CT requirement for chlorine depends on the required log removal, the chlorine concentration, the water temperature and pH. The dependence on the chlorine concentration is because the dilution coefficient n in the Chick-Watson Law not being exactly 1 as often assumed. The dependence on the water temperature and pH is because of the basic chlorine chemistry. Hypochlorous acid, formed after the addition of chlorine gas, calcium hypochlorite (HTH) or sodium hypochlorite (bleach), establishes an equilibrium with the hypochlorite ion: At higher pH, the equilibrium is pushed to the right, leaving more of the chlorine in the hypochlorite form, graphically shown in Figure 6.1.

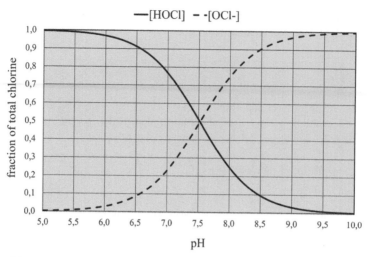

—[HOCl] - -[OCl-]

Figure 6.1: Dissociation of hypochlorous acid

Hypochlorous acid is roughly ten times more efficient at inactivating micro-organisms than the hypochlorite ion, which means that the mixture becomes weaker as the pH is elevated. The CT product required must be higher at higher pH to achieve the same log removal. The dissociation constant is also affected by temperature, which brings the water temperature into play.

The CT product required for the inactivation of *Giardia* by chlorine can be approximated with the temperature in degrees Celsius (Peter Martin, 1993):

$CT_{required} = 0.2828 \times pH^{2.69} \times C^{0.15} \times (\log \text{removal}) \times 0.933^{t\text{-}5}$	Equation 6.4

Example 6.5

Use the Martin approximation to demonstrate how the inactivation of *Giardia* is affected by pH. Calculate CT required for 1.2 mg/L free chlorine to attain log removal of 2 for *Giardia* at 15°C and pH 7.0; and log removal of 3 at 15°C and pH 9.0.

CT product required = 54.5

CT product required = 160.8

An alternative way to determine CT required is to use lookup tables, such as the abbreviated Table 6.3 taken from a large, comprehensive set of tables.

Table 6.3: Required CT values for *Giardia* inactivation at 15°C by free chlorine [a]

Cl$_2$ Residual (mg/L)	pH = 7.0 Log Inactivation			pH = 8.0 Log Inactivation			pH = 9.0 Log Inactivation		
	1	2	3	1	2	3	1	2	3
0.4	23	46	69	33	66	99	47	94	141
0.8	24	48	72	35	70	105	50	100	150
1.2	25	50	75	37	74	111	53	106	159
1.6	26	52	78	39	78	117	56	112	168
2.0	28	56	84	41	82	123	59	118	177

[a] Condensed from Table C-4 in USEPA 1999: C-5

Example 6.6

Read off Table 6.3 to determine the CT required for 1.2 mg/L free chlorine to attain log removal of 2 at pH 7.0; and log removal of 3 at pH 9.0. (These two scenarios are the same as in Example 6.5.)

CT required = 50 (compared to 54.5 obtained with the Martin approximation in Example 6.5)

CT required = 159 (compared to 160.8 obtained with the Martin approximation in Example 6.5)

The tables are quick and easy to use if the lines and columns correspond exactly to the problem at hand. If not, linear interpolation provides the most accurate solution, but it might take up to seven interpolations – a tedious process.

Example 6.7

Compare the Martin approximation with the value interpolated from Table 6.3 for a free chlorine residual of 1.3 mg/L free chlorine to attain log removal of 1.6 for *Giardia* at 15 °C and pH 8.2.

Martin approximation: CT required = 67.6

In Table 6.3, neither the requested log removal, nor the pH and free chlorine residual correspond to the listed values. Follow the accompanying diagram for the recommended sequence of interpolation.

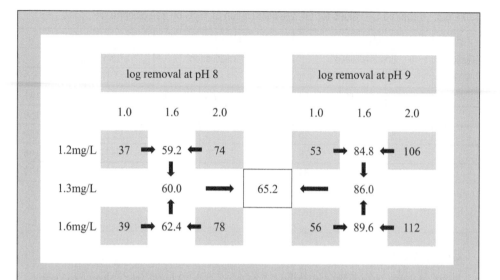

- Copy the four values bracketing the desired value from Table 6.3 for pH 8 and pH 9. These values are shaded in the diagram.

- Interpolate each of the four data pairs to the desired log removal of 1.6, indicated by the short horizontal arrows.

- Interpolate each of the two remaining data pairs to the free chlorine residual of 1.3 mg/L, indicated by the vertical arrows.

- Interpolate between the two remaining data points to find the desired value at pH 8.2, indicated by the long horizontal arrows.

- The final value is a CT product of 65.2 which compares well with the 67.6 from Martın's approximation.

The inactivation of viruses by free chlorine is less complicated, as the required CT only depends on water temperature and pH, shown in Table 6.4.

Table 6.4: Required CT values for virus inactivation by free chlorine at pH 6.0 to pH 9.0 [a]

Inactivation	t = 5°C	t = 10°C	t = 15°C	t = 20°C	t = 25°C
Inactivation = log 2	4.0	3.0	2.0	1.0	1.0
Inactivation = log 3	6.0	4.0	3.0	2.0	1.0
Inactivation = log 4	8.0	6.0	4.0	3.0	2.0

[a] Condensed from Table C-7 in USEPA 1999: C-8.

Example 6.8

Find the CT required for log inactivation = 2.5 of viruses at water temperature of 18°C and pH 8.0 from Table 6.4.

CT required at 15°C and log inactivation of 2.0 = 2.0 (from Table 6.4)

CT required at 15°C and log inactivation of 3.0 =3.0 (from Table 6.4)

CT required at 15°C and log inactivation of 2.5 = 2.5 (interpolate between log 2 and log 3)

CT required at 20°C and log inactivation of 2.0 = 1.0 (from Table 6.4)

CT required at 20°C and log inactivation of 3.0 = 2.0 (from Table 6.4)

CT required at 20°C and log inactivation of 2.5 = 1.5 (interpolate between log 2 and log 3)

CT required at 18°C and log inactivation of 2.5 = 2.1 (interpolate between 15°C and 20°C)

6.4.2. Chloramination

If ammonia is added to water with free chlorine in the right proportion, monochloramine is formed which is also a disinfectant. Although it is not as effective as free chlorine, it has the advantage that it is more stable to maintain a residual in the water for a longer time. Rand Water, for example, uses monochloramine after primary disinfection to maintain sufficient residual of disinfectant over its vast supply area where it may take a week for treated water to reach the furthest points. Although it is not suggested to use monochloramine for primary disinfection, it is included here to allow a comparison of its relative effectiveness with other disinfectants. The CT requirements for inactivating *Giardia* and viruses are shown in Table 6.5 and Table 6.6 respectively.

Table 6.5: Required CT values for *Giardia* inactivation by monochloramine at pH 6.0 to pH 9.0 [a]

Inactivation	t = 5°C	t = 10°C	t = 15°C	t = 20°C	t = 25°C
Inactivation = log 1	735	615	500	370	250
Inactivation = log 2	1470	1230	1000	735	500
Inactivation = log 3	2200	1850	1500	1100	750

[a] Condensed from Table C-10 in USEPA 1999: C-10.

Table 6.6: Required CT values for virus inactivation by monochloramine [a]

Inactivation	t = 5°C	t = 10°C	t = 15°C	t = 20°C	t = 25°C
Inactivation = log 2	857	643	428	321	214
Inactivation = log 3	1423	1067	712	534	356
Inactivation = log 4	1988	1491	994	746	497

[a] Condensed from Table C-11 in USEPA 1999: C-10.

6.4.3. Chlorine dioxide

The CT requirements for inactivating *Giardia* and viruses with chlorine dioxide are shown in Table 6.7, and Table 6.8 respectively.

Table 6.7: Required CT values for *Giardia* inactivation by chlorine dioxide at pH 6.0 to pH 9.0 [a]

Inactivation	t = 5°C	t = 10°C	t = 15°C	t = 20°C	t = 25°C
Inactivation = log 2	8.7	7.7	6.3	5.0	3.7
Inactivation = log 3	17.0	15.0	13.0	10.0	7.3
Inactivation = log 4	26.0	23.0	19.0	15.0	11.0

[a] Condensed from Table C-8 in USEPA 1999: C-9.

Table 6.8: Required CT values for virus inactivation by chlorine dioxide at pH 6.0 to pH 9.0 [a]

Inactivation	t = 5°C	t = 10°C	t = 15°C	t = 20°C	t = 25°C
Inactivation = log 2	5.6	4.2	2.8	2.1	1.4
Inactivation = log 3	17.1	12.8	8.6	6.4	4.3
Inactivation = log 4	33.4	25.1	16.7	12.5	8.4

[a] Condensed from Table C-9 in USEPA 1999: C-9.

6.4.4. Ozone

The CT requirements for inactivating *Giardia* and viruses with ozone are shown in Table 6.9 and Table 6.10 respectively.

Table: 6.9 Required CT values for *Giardia* inactivation by ozone [a]

Inactivation	t = 5°C	t = 10°C	t = 15°C	t = 20°C	t = 25°C
Inactivation = log 2	0.63	0.48	0.32	0.24	0.16
Inactivation = log 3	1.30	0.95	0.63	0.48	0.32
Inactivation = log 4	1.90	1.43	0.95	0.72	0.48

[a] Condensed from Table C-12 in USEPA 1999: C-11

Table 6.10: Required CT values for virus inactivation by ozone [a]

Inactivation	t = 5°C	t = 10°C	t = 15°C	t = 20°C	t = 25°C
Inactivation = log 2	0.60	0.50	0.30	0.25	0.15
Inactivation = log 3	0.90	0.80	0.50	0.40	0.25
Inactivation = log 4	1.20	1.00	0.60	0.50	0.30

[a] Condensed from Table C-13 in USEPA 1999: C-11.

6.4.5. Comparison of required CT values

Using the CT requirements from Table 6.3 to Table 6.10, a comparison of the relative effectiveness of different disinfectants is presented in Table 6.11.

Table 6.11: Comparison of required CT values at water temperature of 15°C and pH 8

Disinfectant	Giardia (3 Log Removal)	Giardia (2 Log Removal)	Viruses (4 Log Removal)
Free chlorine at 2 mg/L and pH 8	123	81	4.0
Monochloramine	1500	1000	994
Chlorine dioxide	13.0	6.3	16.7
Ozone	0.63	0.32	0.60

Several points emerge from Table 6.11:

- The 3-log removal of *Giardia* and *Cryptosporidium* requires higher CT values than the 4-log removal of viruses. For surface water compromised by protozoan contamination, the virus removal requirement is therefore automatically met if the CT removal requirement of *Giardia* and *Cryptosporidium* are met. For the disinfection of groundwater which is not influenced by surface water, the disinfection parameters are determined by virus inactivation.
- Free chlorine, compared to monochloramine, is about 10 times more effective against *Giardia* and 100 times more effective against viruses.
- The pH has a strong influence on the effectiveness of free chlorine, but not on the other disinfectants provided the pH is 9 or below.

Table 6.3 to Table 6.10 show throughout that disinfection is less effective in cold water than in warm water. When performing CT calculations, we must always assume the worst-case scenario. In other words, we must assume the:

- lowest water temperature;
- highest water pH anticipated;
- lowest chlorine residual; and
- the shortest contact time.

For more pertinent information on disinfectants used in South Africa, a useful manual had been made available by the Water Research Commission which provides further practical considerations regarding disinfection (Van der Walt et al., 2009).

6.5. DETERMINATION OF ACTUAL CONCENTRATION C ACHIEVED

Effective disinfection must be guaranteed at all times. The concentration used in CT calculations must, therefore, be the lowest disinfection residual that can be foreseen at the closest customer. The practice of taking one grab sample per shift or per day is therefore not completely satisfactory, as there may be certain times of the day when the actual concentration may be less than indicated by the grab sample owing to flow variations, faulty or inaccurate chemical dosing, short lapses in upstream treatment (such as periods of filter ripening), etc. The tragic incident in 2000 in Walkerton, Canada when five persons died and 2 500 fell ill because of improper chlorine dosing and monitoring of chlorine residuals, vividly illustrated the shortcomings of intermittent grab sampling and operator indifference. (The two operators spent one year in prison and nine months in house arrest respectively.) One of the recommendations of the Commission of Enquiry was:

> All water should be continuously monitored, with alarms and automatic shut-off systems if something goes wrong. All municipal providers should have an adequate sampling plan and samples should be taken at vulnerable times, such as after a flood or heavy rainfall. All testing should be done at laboratories accredited by the Ministry of the Environment.

Continuous monitors for disinfectant residual will show the lowest concentration during the interval considered, which should be used to calculate the CT product achieved.

6.6. DETERMINATION OF ACTUAL CONTACT TIME T

Water treatment plants and their subsequent distribution systems consist of combinations of pipes and tanks. In both cases, the average residence time T is calculated with Equation 6.5:

$$T = \frac{V}{Q}$$	Equation 6.5
T = average residence time (h) V = water volume in reactor (m³) Q = flow rate (m³/h)	

Example 6.9

Determine the average residence time in a pipeline with diameter 600 mm and length 3700 m; and a tank with footprint 12 m x 30 m filled to a depth of 3.4 m. The flow rate = 420 m³/h = 7 m³/min in both cases.

Volume = π x 0.6²/4 x 3700 = 1046 m³, therefore T = 1046 / 7 = 149 minutes

Volume = 12 x 30 x 3.4 = 1224 m³, therefore T = 1224 / 7 = 175 minutes

---

We are, however, not so much interested in the average residence time, but rather in the minimum residence time to ensure that all the organisms had adequate contact with the disinfectant. The difference between the average and minimum residence time is owing to short-circuiting when some water follows a short flow path to the exit, and other water ends up in semi-stagnant zones. If there would be no short-circuiting, then the average and minimum residence times would be the same.

6.6.1. Calculating T for pipelines

There is negligible short-circuiting (or longitudinal dispersion when we consider linear flow) in pipelines because of the low possibility that some water could "overtake" the rest while flowing through the pipe. It is, therefore, customary to assume plug flow for pipes and to calculate the minimum residence time with equation 6.5.

6.6.2. Calculating T for tanks

For tanks, short-circuiting is unavoidable. To calculate the minimum residence time, Equation 6.5 is modified by adding a "baffling factor" or "short-circuiting factor" F_{sc}:

$$T = F_{SC} \times \frac{V}{Q}$$	Equation 6.6

For no short-circuiting (as in pipes), $F_{sc} \approx 1$; for a high degree of short-circuiting, F_{sc} would be close to zero. Consider the four tanks shown in Figure 6.2.

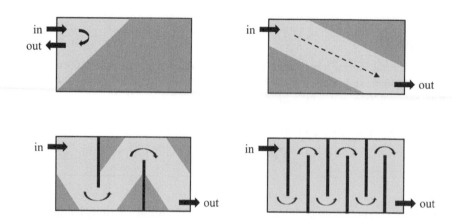

Figure 6.2: Schematic top views of tanks with different degrees of short-circuiting, stagnant zones depicted darker

An interpretation of Figure 6.2 is provided in Table 6.12.

Table 6.12: Typical tank layouts with appropriate baffling factors [a]

Baffling Condition	T_{10}/T Ratio	Description
Unbaffled (backmix)	0.1	No baffles, very low L/W ratio, high inlet / outlet velocities
Poor	0.3	Unbaffled inlets and outlets, no intra-basin baffles
Average	0.5	Baffled inlets / outlets with some intra-basin baffles
Superior	0.7	Perforated inlet baffle, intra-basin serpentine flow path
Perfect (plug flow)	1.0	Pipe flow, perforated inlets / outlets/ intra-basin baffles

[a] Taken from Table 3 of Rush (2002).

Figure 6.2 and Table 6.12 have educational value, but normally not accurate enough for the real world. The most accurate method is to perform a tracer study (only possible for existing tanks) or by computational fluid dynamic (CFD) modelling. If either of these methodologies are available, the minimum residence time is taken as T_{10}, which is the time required for 10% of the tracer to wash out of the tank.

A tracer study can be done by adding a pulse input of tracer at the tank inlet, and then monitor the concentration of the tracer in the water at the outlet. When the tracer concentration is plotted against time, the following is observed:

- Without short-circuiting, the tracer pulse would appear as a narrow Gaussian curve indicating plug flow.

- The more short-circuiting, the more the curve is stretched out as the tracer makes its way to the outlet.

- With stagnant spots in the tank, a very long "tail" is measured.

A further complication with tanks arises when estimating the minimum water depth. For maximum smoothening out of flow variations, one would wish to draw the tank down to the very bottom in extreme cases. To maximise residence time, one would wish to keep the tank at full water depth at all time. For tanks that have a dual purpose of buffering as well as contact, a compromise must be found by setting a minimum allowable water level in the tank and to guarantee that level by providing alarms or automatic shutdown when it is reached. This minimum level then becomes the value to be used in calculating the minimum water volume and retention time.

Example 6.10

Calculate the minimum residence time in a tank with footprint 12 m x 30 m filled to a minimum level of 60% of the total depth of 3.4 m if the flow rate is 420 m³/h. A previous tracer test indicated a short-circuiting factor of 0.56.

Volume of tank when 100% full = 12 x 30 x 3.4 = 1224 m³

Volume of tank when 60% full = 0.60 x 1224 = 734.4 m³

Flow rate = 420 / 60 = 7 m³/minute

Minimum residence time T = 0.56 x 734.4 / 7 = 59 minutes

6.6.3. Combinations of tanks and pipes

Long pipelines provide long residence time (roughly 20 minutes for every 1 km of pipeline). They are often utilised as contact reactors in combination with storage and other tanks along the way. The proper approach in these cases is to calculate C and T for every part of the total system, and then to add the CT products.

Example 6.11

The tank in Example 6.10 feeds a pipeline with diameter 600 mm. The distance from the tank to the first customer is 2400 m. The monitoring of chlorine residuals indicates a minimum concentration of 2.4 mg/L at the tank outlet, and 1.3 mg/L at the connection to the first customer. Estimate the CT product for the system.

Flow rate = 420 / 60 = 7 m³/minute

Volume of pipeline = π x 0.6²/4 x 2400 = 679 m³

Residence time in pipeline = 679 / 7 = 97 minutes

CT for pipeline = 1.3 x 97 = 126 mg.min/L

CT for tank = 2.4 x 59 (Example 6.10) = 142 mg.min/L

Total CT for system = 126 + 142 = 268 mg.min/L

6.7.   REMEDIAL ACTIONS

What to do if the actual CT product is smaller than required? The first impulse would be to increase the disinfectant dosage, which is the only short-term solution. In the longer term, however, a more efficient remedy may be to:

- Try to spread the hydraulic load more evenly over the day to prevent sharp flow peaks for short periods of the day. High flow rates translate to lower residence times in all hydraulic units.

- Make changes to the upstream physical treatment to reduce the log removal burden on the disinfection process.

- Adjust the minimum water levels in the tanks upwards to increase the residence time.

6.8.   CHLORINE DEMAND

When chlorine is added to water, whether in the form of free chlorine or monochloramine, it is inevitable that its concentration will diminish with time:

- In the case of monochloramine, it is slowly consumed as it reacts with other compounds in the water.

- In the case of free chlorine, it first forms monochloramine which must be reacted away with more free chlorine. This gets us to the "breakpoint", a concept important for swimming pool management covered in Chapter 10. Even more chlorine must be added beyond the breakpoint to get a free chlorine residual.

All the material presented thus far in the chapter dealt with the residual concentration C. But how much chlorine must be added by an operator to get a desired residual? It is useful to simplify the reaction of chlorine with water as two consecutive steps – an initial fast reaction followed by a slow reaction.

The fast reaction is important in the presence of contaminants which react rapidly with chlorine, such as ammonia. This can be measured quickly and simply with a water sample dosed with a known concentration of free chlorine, kept in the dark and measuring the residual chlorine after say 20 to 30 minutes.

The slow reaction causes a loss of chlorine as it is broken down by photolysis or as it reacts with organic compounds or new contaminants that may have entered the distribution system – processes that continue for hours or days until it reaches the consumers. In these cases, we also need to be able to measure the rate of decay. A simple approach is to assume first-order kinetics for the chlorine decay, given by Equation 6.7 in differential form:

$$\frac{dC}{dt} = -k \cdot C$$	Equation 6.7

Equation 6.7, after integration, leads to:

$$C = C_0 \cdot e^{-kT}$$

| Equation 6.8 |

C is the concentration of chlorine remaining T hours after an initial dose of C_0, with k the bulk decay coefficient. The bulk decay constant is measured by dosing a water sample with chlorine, measuring the residual chlorine after 30 minutes (which provides the chlorine concentration after the initial fast reaction) and again after 24 hours in the dark (the additional chlorine lost by the slow reactions).

Example 6.12

A water sample is dosed with 4.0 mg/L and placed in the dark. After 30 minutes, the concentration dropped to 2.4 mg/L and after 24h to 1.5 mg/L. Calculate the chlorine loss by initial fast reaction and the bulk decay coefficient for free chlorine.

Chlorine lost by fast reaction = 4.0 – 2.4 = 1.6 mg/L

Chlorine before slow reaction = 2.4 mg/L; after slow reaction = 1.5 mg/L

Slow reaction time = 24 – 0.5 = 23.5 h

Bulk decay coefficient = 0.0200 /h

Example 6.13

The water of Example 6.12 enters a distribution system which consists of a pipeline and a reservoir with a combined retention time of 2.2 days. Estimate the initial dose of free chlorine at the treatment plant to ensure a free chlorine residual of 0.8 mg/L at the outlet of the reservoir.

Required chlorine after slow reaction = 0.8 mg/L

Slow reaction time = 2.2 days = 2.2 x 24 = 52.8 h

Required chlorine before slow reaction = 2.3 mg/L

Chlorine lost by fast reaction = 1.6 mg/L (from Example 6.12)

Required chlorine dose at treatment plant = 1.6 + 2.3 = 3.9 mg/L (Equation 6.8)

Earlier research in South Africa showed that the bulk decay of both monochloramine and free chlorine does not necessarily follow first-order kinetics. For more detailed work, the decay kinetics is better described with an nth-order equation. Also, the insides of the pipes, after many years, develop a biofilm which also exerts additional chlorine demand. For more details on a comprehensive methodology for measuring and modelling chlorine decay, the reader is referred to the work of Viljoen et al. (1997).

6.9. REFERENCES

Martin, P. 1993. "Calculating CT Compliance." *Journal of the American Water Works Association*, 85(12): 12.

New South Wales. 2015. Indicators, Reference Pathogens and Log Reductions. New South Wales Recycled Water Information Sheet Number 2. https://www.industry.nsw.gov.au/__data/assets/pdf_file/0010/180568/IS2_Surrogates-Ref-Pathogens-LRVs.pdf

Rush, Brock. 2002. CT Disinfection Made Simple. Calgary: Alberta Environment. http://www.water-research.net/Waterlibrary/CT_LookupTable/21%20CTMadeSimple.pdf

United States Environmental Protection Agency. 1999. Disinfection Profiling and Benchmarking Guidance Manual. Report 815-R-99-013. https://nepis.epa.gov/Exe/ZyPDF.cgi/20002249.PDF?Dockey=20002249.PDF

United States Environmental Protection Agency. N.d. Surface Water Treatment Rule Fact Sheet. https://www.epa.gov/sites/production/files/documents/SWTR_Fact_Sheet.pdf

Van der Walt, M., Kruger, M., and Van der Walt, C. 2009. The South African Oxidation and Disinfection Manual. Report TT406/09. Pretoria: Water Research Commission. http://www.wrc.org.za/wp-content/uploads/mdocs/TT%20406%20web%20Industrial%20Water%20Mangement.pdf

Viljoen, O.J., Haarhoff, J. and Joubert, J.H.B. 1997. *The Prediction for Chlorine Decay from Potable Water in Pipeline Systems*. Report 704/1/97. Pretoria: Water Research Commission.

ADSORPTION

7.1. INTRODUCTION

As raw water quality in South Africa deteriorates and demand increases, we are forced to turn to options such as wastewater reclamation along with newer, more advanced treatment processes. It is quite clear that treatment process trains of the future will increasingly rely, in some way or other, on the process of adsorption. Although not common at South African municipal treatment plants at present, there can be little doubt that a good understanding of the process will be indispensable in the future.

7.2. ACTIVATED CARBON

Activated carbon (AC) is the most common adsorbent used in drinking water treatment. It can be manufactured from different organic base materials. The different base materials, coupled with different technologies for manufacture and activation, lead to AC products with significant differences. To be able to compare the different products on a quantitative basis, numerous empirical measures had been developed to characterise the products. Besides physical parameters such as particle size, shape, porosity and density, some parameters strive to indicate their adsorptive capacity such as the iodine, methylene blue or tannin numbers. These values would be an indication of the type/size molecules that could be adsorbed and the respective values roughly correspond to the internal pore size distribution. The difficulty lies in the fact that the adsorptive capacity does not only depend on the AC, but also on the contaminant to be removed, general water quality, pH and ion content and competitive adsorption of other compounds. A South African study a few years ago showed that these commonly used "adsorption numbers" do not predict the ability of AC to adsorb the odour-producing geosmin or 2-MIB in the water from Vaal Dam with sufficient precision (Linde et al., 2003). If AC is considered for practical application, further laboratory studies are necessary.

The manufacture of AC starts with a carbonaceous base material such as bamboo, coconut husk, peat, wood, lignite or coal. The base material is subjected to two consecutive steps. The material is first carbonised by heating it to roughly 800°C in the presence of an inert gas such as argon or nitrogen. It is then activated by replacing the inert gas with an oxidising gas such as steam or oxygen at about 1000°C. In some cases, the processes are modified by adding certain chemicals such as acid or strong base at the start, which allows the subsequent carbonisation and activation at lower temperatures and shorter reaction times.

AC has a porous structure, as some of the base material components are "burned out" by the carbonisation and activation steps. As a result, the density of activated carbon is lower than the base material – typically in the range 300 to 600 kg/m³. The internal pores within the AC (see Figure 7.1) are classified as micro- (smallest), meso- or macropores (largest). The internal pores increase the surface area of each grain tremendously – the surface areas of AC products are in the range 1000 to 3000 m²/g of material. Compare that to the surface area of 300 m²/g for commercial cement used for construction, which is an extremely fine powder.

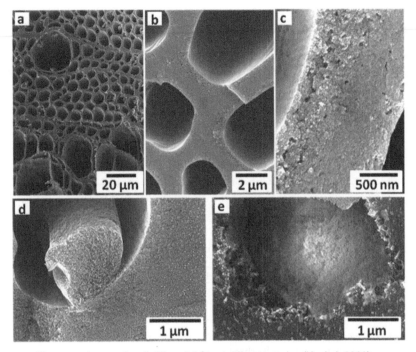

Figure 7.1: Electron micrographs of different AC's at different scales (Kaskel, 2020)

7.3. ADSORPTION CAPACITY AT EQUILIBRIUM

AC testing starts with the determination of the adsorption capacity of the product at equilibrium conditions. Different amounts of AC are added to water samples containing the problem contaminant. After allowing enough reaction time to reach equilibrium, with controlled agitation at controlled temperature, the contaminant concentration remaining in the water is measured. These studies normally start with preliminary tests to determine the experimental conditions to ensure equilibrium – a good example of such a systematic investigation is offered, for example, by Ashgar et al., (2015).

Although the results obtained could be useful on their own, much value is added to the experiment by fitting the results to a mathematical model, which provides the opportunity to extrapolate the results to other scenarios. Two such models, in the case of adsorption called isotherms, are the commonly used Langmuir and Freundlich isotherms. Both are underpinned by rational assumptions, not presented here.

A key parameter in both the Langmuir and Freundlich isotherms is the contaminant loading q_e on the AC at equilibrium, expressed as the mass of contaminant removed divided by the mass of AC.

Example 7.1

The concentration of a contaminant is reduced from 120 µg/L to 44 µg/L after 24 mg/L of AC is added. Calculate the mass loading q_e of the contaminant on the AC.

C removed $= 120 - 44 = 76$ µg/L

AC added $= 24$ mg/L

Mass loading $q_e = 76 / 24 = 3.17$ µg/mg

7.3.1. The Langmuir isotherm

The Langmuir isotherm is usually presented in the form of Equation 7.1:

$$q_e = \frac{q_{max} \cdot b \cdot C_e}{1+b \cdot C_e}$$	Equation 7.1
q_e = mass adsorbate / mass adsorbent q_{max} = constant relating to the maximum value of q_e b = constant relating to the strength of adsorption C_e = concentration of solution at equilibrium	

To determine the Langmuir constants q_{max} and b from experimental data, Equation 7.1 is linearised:

$$\frac{1}{q_e} = \frac{1}{q_{max} \cdot b \cdot C_e} + \frac{1}{q_{max}}$$	Equation 7.2

To determine the Langmuir constants, calculate the mass loading q_e for every testing jar. Then plot $1/q_e$ on the y-axis against $1/C_e$ on the x-axis using linear scales and find the best-fit straight line through the data points. The reciprocal of the intercept provides q_{max}. The reciprocal of the slope of the regression line, together with q_{max}, allows the calculation of b.

With the Langmuir constants known, it is convenient to introduce the AC dosage concentration m (Equation 7.3) and the removal fraction x of the starting concentration C_o (Equation 7.4):

$$q_e = \frac{x \cdot C_0}{m}$$	Equation 7.3

$$C_e = (1-x) \cdot C_0$$	Equation 7.4

The Langmuir isotherm is now rewritten as Equation 7.5 to calculate the required dosage concentration m to remove a desired fraction x directly:

$$m = \frac{x \cdot [1 + b \cdot (1-x) \cdot C_0]}{q_{max} \cdot b \cdot (1-x)}$$	Equation 7.5

7.3.2. The Freundlich isotherm

Equation 7.6 provides the usual form of the Freundlich isotherm:

$$q_e = K \cdot C_e^{1/n}$$	Equation 7.6
q_e = mass adsorbate / mass adsorbent K = Freundlich constant n = Freundlich constant C_e = concentration of solution at equilibrium	

To determine the Freundlich constants K and n from experimental data, the equation is linearised:

$$\log q_e = \log K + \frac{1}{n} \cdot \log C_e$$	Equation 7.7

To determine the Freundlich constants, prepare the data as for the Langmuir isotherm. Now plot log q_e on the y-axis against log C_e on the x-axis using linear scales and find the best-fit straight line through the data points. Use the intercept on the y-axis to find K. The reciprocal of the slope in the value of n.

Using the same substitutions as in the case of Langmuir, the Freundlich isotherm is now rewritten as Equation 7.8 to provide the dosage m required for a desired removal fraction x directly:

$$m = \frac{x \cdot C_0}{K \cdot [(1-x) \cdot C_0]^{1/n}}$$

Equation 7.8

7.3.3. Isotherm application

It is useful to compare some isotherm data from the literature. Two randomly picked references presented isotherm constants shown in Table 7.1. These constants were used to calculate a range of possible removal scenarios in Figure 7.2, which demonstrates the power of presenting the data in isotherm format. Once the isotherm constants are determined, a whole range of different dosages and removal ratios can be explored. From Figure 7.2, for example, we can find that the dosage for 40% removal of geosmin should be 45 mg/L and that it should be increased to 95 mg/L if we wish to increase the removal to 80%. In addition, that it would require about 15 mg/L of AC to remove 70% of chloroform – the major constituent of trihalomethanes.

Table 7.1: Isotherm constants from the literature [a]

	Constant 1	Constant 2	Comment
Geosmin [b]	$Q_{max} = 0.90$	b = 0.64	For Langmuir isotherm
2-Methylisoborneol [b]	$Q_{max} = 0.80$	b = 0.62	For Langmuir isotherm
Chloroform [c]	K = 2.6	n = 1.37	For Freundlich isotherm
Bromoform [c]	K = 20	n = 1.92	For Freundlich isotherm

[a] Ashgar et al 2015 and Snoeyink 1990; [b] For concentrations expressed in ng/L; [c] For concentrations expressed in mg/L

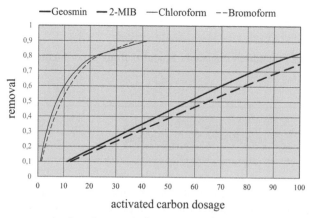

Figure 7.2: Isotherm examples for odour-producing compounds geosmin and 2-MIB and trihalomethanes chloroform and bromoform

Despite some shortcomings, isotherms are great tools for predicting the theoretical maximum that AC can remove under equilibrium conditions with simple laboratory procedures. When doing the isotherm tests, the contact times must be in the order of hours or even days to ensure that the total capacity of the AC for adsorbing the contaminant is utilised.

Isotherms do not, however, tell the full story. The process of adsorption is not instantaneous, as it takes time for the contaminant to reach the surface of the AC grains, then some more time to diffuse into their macro-, meso- and micropores. Treatment plants do not allow such lengthy contact times. It is therefore usually necessary to supplement the isotherms with the quantification of the kinetic constraints to adsorption. The kinetic constraints depend on the type of reactor used for contacting the water with the carbon, discussed next.

7.4. ADSORPTION SYSTEMS

AC is most commonly available in either powdered (PAC) or granular (GAC) form. Much research is currently ongoing to "package" and "enhance" AC in different shapes and forms such as carbon nanotubes and doped activated carbon, but they do not offer competitive options for large-scale drinking water treatment yet and therefore not discussed here. In a nutshell, the choice between PAC and GAC hinges on several technical considerations:

- Raw water quality fluctuations. If the raw water quality is subject to frequent swings, only requiring adsorption intermittently, it is best to use PAC, which can be quickly brought on-line when required and terminated when not.
- Initial capital cost. The contacting system for PAC is considerably cheaper.
- Storage and handling. From an operational point of view, the daily handling of a fine, black, dusty powder like PAC is more troublesome than GAC, which is only handled at intervals in wet form. On the other hand, PAC handling is limited to getting it into the water only – from there it becomes part of the treatment residue along with sludge and wash water while GAC must be periodically removed and replaced with special effort and equipment.
- Dosage control. The dosage of PAC can be precisely controlled according to the raw water quality while GAC allows no direct dosing control.
- Reactivation. GAC offers the opportunity to be reactivated to save costs – PAC not.

The methods of contacting the AC with water are dependent on whether PAC of GAC is used, discussed under the headings following.

7.4.1. PAC mixing tanks

PAC is delivered to site as a dry, fine powder in sealed bags. When PAC dosing is required, it is added to water in a batch mixing tank and stirred to prepare a PAC slurry of known concentration using specialised wetting equipment. From here, the slurry is pumped or flows to a dosing point to deliver the desired PAC dosage.

The effectiveness of the PAC depends firstly on the contact time allowed between the dosing point and the point where the PAC is separated. Conventionally, PAC removal takes place either during pre-filtration separation (sedimentation or dissolved air flotation), or during filtration. The minimum contact time (which should be tested for each case) is usually between 10 and 20 minutes for good adsorption.

The effectiveness of PAC also depends on the hydraulic efficiency of the contact reactor. The differences between backmix and plugflow reactors were discussed in Chapter 5 which dealt with the efficiency of disinfection. Analogously, PAC performs best in a plugflow reactor. Treatment plants therefore utilise either pipelines or long zig-zag channels where the turbulence induced by the changes in direction provide the mixing energy needed to keep the PAC in suspension. Both pipes and zigzag channels provide near plugflow conditions.

7.4.2. GAC contact columns and filters

The use of GAC requires much more sophisticated infrastructure. A bed of GAC must be placed in a reactor, and the water then allowed to drain through the bed at a precisely controlled flow rate. At frequent intervals (say weekly), the GAC beds must be backwashed to remove the particles trapped among the GAC grains to restore the bed to its original cleanliness. However, after a longer period (say six months to a year), the adsorptive capacity of the carbon will be exhausted and the GAC must then be physically removed and replaced with fresh GAC.

Two types of GAC contact reactors are prevalent. The first is a rectangular filter box of concrete, similar to the filters used for sand filtration – here the GAC is spread in a layer about 1 m deep over a fairly large area. The second is a tall, vertical cylinder of steel – here the GAC could be 2 m to 4 m deep, resulting in a smaller footprint. These geometrical variations make little difference to the adsorption performance, as the determining factor is the contact time, just as for PAC. The contact time for GAC columns and filters is expressed as the Empty Bed Contact Time (EBCT):

$$EBCT = \frac{A \cdot d}{Q}$$	Equation 7.9
EBCT = empty bed contact time (minutes) A = surface area of GAC bed (m^2) D = depth of media bed (m) Q = water flow rate (m^3/min)	

Example 7.2

Calculate the EBCT for a treatment plant producing 5.4 m^3/min for two treatment plants which use GAC in a filter arrangement and a columns arrangement respectively. In the first, there are three filters each 4 m x 6 m with GAC beds 900 mm deep. In the second, there are 10 columns with diameter of 1.5 m and depth of 4 m.

Filters: A = 3 x 4 x 6 = 72 m²; d = 0.9 m

EBCT = 72 x 0.9 / 5.4 = 12.0 minutes

Columns: A = 10 x π x 1.5² / 4 = 17.7 m²; d = 4 m

EBCT = 17.7 x 4.0 / 5.4 = 13.1 minutes

How do we know when to replace the GAC? In the usual case where water flows downwards through the GAC bed (it could also be directed upwards to hold the GAC in a fluidised bed, not discussed here), the topmost GAC of a fresh bed will remove the contaminant (the adsorption zone), while the GAC at the bottom retains its full adsorptive capacity. After a while, the top GAC reaches full saturation and the removal zone moves further down. Refer to Figure 7.3, where dark grey indicates fully saturated GAC, light grey the adsorption zone which is partly saturated GAC, and white GAC with no saturation. The adsorption zone moves downward with time until it reaches the bottom of the bed. At this point, the contaminant concentration in the effluent shoots up to indicate a breakthrough condition. Soon thereafter, if the GAC bed is not replaced, all removal will cease with a fully saturated bed.

For the removal of single target contaminants, the detection of the breakthrough point is simple. The effluent should be frequently monitored for the target contaminant and the GAC bed replaced when the contaminant concentration reaches a predetermined threshold. For multi-component mixtures such as organic carbon, an indicator parameter should be used which is preferably quick and easy to use. At the Rietvlei Water Treatment Plant near the City of Tshwane, for example, the UV absorbance at 254 nm is used as a routine parameter. Water contains numerous compounds that would compete for adsorption sites on the AC. It is possible that those which are "loosely" attached are displaced by compounds with superior adsorptive proprieties which may lead to the early breakthrough of some undesirable compounds.

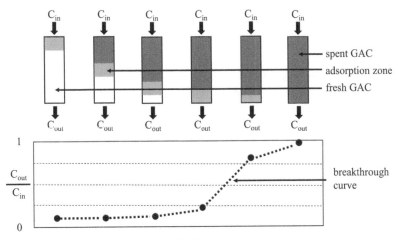

Figure 7.3: Schematic of breakthrough in a GAC column

The prediction of the point of breakthrough is a key parameter for the economic viability, as well as the design of GAC columns of filters. For this reason, laboratory investigations do not only include equilibrium tests (discussed in Section 7.3), but also column tests to assess the kinetics of GAC adsorption. Small glass columns with diameter of typically less than 50 mm is used at first. If the results are promising, the next step would be pilot filters of say 100 mm or more to verify the GAC performance at more realistic scale. For direct comparison across laboratory, pilot and plant scales, the capacity of the GAC is expressed as bed volumes:

$$\text{bed volumes } = \frac{\text{volume treated}}{A \cdot d}$$	Equation 7.10

volume treated = volume before breakthrough (-)

A = area of GAC bed (m²)

d = depth of GAC bed (m)

Example 7.3

A small laboratory filter has a diameter of 40 mm and holds a GAC bed of 180 mm deep. A volume of 47 L passes through the bed before breakthrough is detected. Calculate the GAC adsorptive capacity in terms of bed volumes.

Volume treated = 0.047 m³ ; d = 0.180 m

Diameter = 0.040 m, therefore A = 0.00126 m²

Bed volumes = 0.047 / (0.00126 x 0.040) = 208

Example 7.4

Use the results of Example 7.2 to estimate the water volume required to perform a GAC pilot test if the diameter of the pilot filter is 135 mm and the GAC depth 450 mm. Allow an extra 50% to ensure that the pilot experiment will be able to run to breakthrough.

Bed volumes = 208 (from Example 7.2); depth d = 0.450 m

Diameter = 0.135 m, therefore A = 0.0143 m²

Water volume required = 1.338 m³ plus 50% = 2.008 m³

Once a treatment plant is in full operation, the bed volume concept remains useful as a first guide to anticipate when breakthrough may be expected. If there is a change in the number of bed volumes before breakthrough over time, it would indicate either changes in the quality of the GAC, or changes in the raw water quality.

7.4.3 Biological Active Carbon (BAC)

In some instances, after GAC has been in use for some time, microbiological activity on the GAC would assist in transforming and removing adsorbed compounds and thereby extending the life of the GAC before regeneration is required. The use of ozone prior to GAC treatment will enhance the breakdown of organic compounds and stimulate microbiological growth.

7.5. REACTIVATION OF GAC

GAC is relatively costly. It is therefore an attractive option to regenerate the GAC after use to restore it to its original adsorption capacity. The most common regeneration technique is thermal reactivation, which generally follows three steps:

- Adsorbent drying at approximately 105°C;
- High temperature desorption and decomposition (500 to 900°C) under an inert atmosphere; and
- Residual organic gasification by a non-oxidising gas (steam or carbon dioxide) at elevated temperatures (800°C).

The heat treatment stage results in desorption, partial cracking and polymerisation of the adsorbed organics. The final step aims to remove charred organic residue formed in the porous structure in the previous stage and re-expose the porous carbon structure to regain its original surface characteristics.

Every time GAC is regenerated, between 5% and 15% of the GAC is burnt off resulting in a loss of adsorptive capacity. Care must be taken to control reactivation such that minimal damage is caused to the pore structure and distribution so that the GAC would retain its adsorptive capacity. Thermal regeneration requires high temperatures making it both an energetically and commercially expensive process. Treatment facilities that rely on thermal regeneration of activated carbon must have a certain minimum degree of utilisation before it is economically viable to have regeneration facilities onsite. As a result, it is common for smaller waste treatment sites to ship their GAC to specialised commercial facilities for regeneration.

7.6. REFERENCES

Asghar, A., Khan, Z., Maqbool, N., Qazi, I.A., and Awan, M.A. 2015. "Comparison of Adsorption Capacity of Activated Carbon and Metal Doped TiO2 for Geosmin and 2-MIB Removal from Water." *Journal of Nanomaterials*, 2020:11p. http://dx.doi.org/10.1155/2015/479103.

Kaskel, Stefan. 2020. Electron micrographs of different activated carbons at different scales. Supplied by Prof. Kaskel at the Technical University Dresden.

Linde, J.J., Freese, S.D. and Pieterse, S. 2003. Evaluation of Powdered Activated Carbon (PAC) for the Removal of Taste and Odour Causing Compounds from Water and the Relationship Between this Phenomenon and the Physico-Chemical Properties of the PAC and the Role of Water Quality. Report 1124/1/03, Pretoria: Water Research Commission. http://www.wrc.org.za/Knowledge%20Hub%20Documents/Research%20Reports/1124-1-03.pdf

Snoeyink, V.L. 1990. Adsorption of Organic Compounds. In *Water Quality and Treatment*, 4th Edition. Denver: American Water Works Association.

TREATMENT RESIDUE

8.1. INTRODUCTION

The main concern of water users and water providers is to ensure enough water of good quality. A secondary concern is the problem of managing the materials removed from the water – a problem that has serious technical, economic and environmental dimensions. At this point in time, there is no clear consensus on how this problem should be dealt with, as there has been a huge shift in policies, practices and environmental constraints over the past two or three decades. This chapter raises these issues in a general, introductory fashion to alert future practitioners to the main concerns, rather than providing detailed technical guidance. Also, it focuses primarily on particulate residue, which is the main problem confronting current treatment plants. An emerging problem, owing to its relatively recent arrival, is the liquid waste produced by membrane filtration. This is addressed cursorily at the end of the chapter. Most of the general principles discussed in the context in particulate residue also apply to liquid residue.

It is useful to structure this chapter according to five conceptual steps, applicable to all waste streams (after Doe, 1990):

- Separation of the residue by different treatment units;
- Collection of the residue from different treatment units;
- Thickening of the residue to decrease its volume and for water recovery;
- Treatment of the residue for partial recovery or immobilisation; and
- Disposal of the remaining residue

A figure often quoted, although it pertains to domestic wastewater treatment, is worth repeating. About 50% of the capital cost of a wastewater treatment is sunk into the treatment of the solids content of the wastewater, although the dry solids content of domestic wastewater is less than 0.03%. For drinking water treatment plants, the capital cost of residue management is not so dramatic yet but is certainly disproportionate to the residue concentration — and certain to rise in the future.

8.2. ESTIMATION OF RESIDUE QUANTITIES

The sizing of a residue handling system depends on two critical parameters. The first is the average mass of residue generated by the water treatment plant, which determines the site requirements for the ultimate disposal of the residue over the long-term. The second is the peak discharge rate of the residue, which determines the hydraulic capacity and the capital cost of the residue processing system before ultimate disposal. It is best to estimate the average and peak rates from long-term treatment plant records if available. The residue load consists of the natural particles in the raw water, as well as particles added by chemical treatment. Good records of both raw water quality and treatment chemicals added are therefore required.

The suspended solids (SS) of the raw water are a direct measure of the mass of residue carried by the raw water. SS is however not measured as a routine parameter. The next best estimate is obtained from turbidity, which is routinely measured. Measurements at numerous South African treatment plants, including both eutrophic and silt-laden raw waters, indicated a relationship always bracketed by 1 NTU $\approx$ 0.5 to 2.0 mg SS/L, approximated to 1 NTU = 1 mg SS/L for raw water turbidity above 100 NTU.

Non-precipitation treatment chemicals are easy to account for:

- Particulate chemicals such as bentonite clay added for better sedimentation of eutrophic water or powdered activated carbon for odour control, all end up directly as residue.
- Organic polymers, although they are added in liquid form, attach to the particles and are therefore removed with the residue. The dry mass equivalent of liquid polymer depends on the specific product used but a first estimate is provided by 1 mg product $\approx$ 0.5 mg dry solids.
- The pathway of other non-precipitating chemicals such as chlorine do not enter the residue stream and are anyway added towards the end of treatment after most residue have already been removed.

The residue mass added by aluminium and iron-based coagulants is controlled by the stoichiometry of the precipitation reactions:

Stoichiometrically, 1.00g Fe^{+3} produces 1.93g $Fe(OH)_3$ and 1.00g Al^{+3} produces 2.89g $Al(OH)_3$.

Example 8.1

Estimate the residue contained in a flow of 15 000 m^3/h raw water with 80 NTU after dosing with 24 mg/L of $Al_2(SO_4)_3.14H_2O$.

Molecular mass $Al_2(SO_4)_3.14H_2O$ = 54 + 288 + 252 = 594 g/mol

Molecular mass Al = 27.0 g/mol

24 mg/L $Al_2(SO_4)_3.14H_2O$ contains = 24 x 2 x 27 / 594 = 2.18 mg/L of Al^{+3}

Residue from raw water = 80 x 1.0 = 80 mg/L

Precipitate from coagulant = 2.18 x 2.89 = 6.3 mg/L

Total residue concentration = 80 + 6 = 86 mg/L = 86 g/m^3

Rate of residue production = 15000 x 86 = 1290 kg/h or 30960 kg/day of dry solids

---

Estimates of the precipitates formed when the pH is raised by the addition of $Ca(OH)_2$ (a strategy used to precipitate heavy metals or to soften the water) are more complex and cannot be tackled with simple stoichiometry as it depends not only on the $Ca(OH)_2$ dosage, but also the target pH and the raw water quality. For $Ca(OH)_2$ addition, it is better to conduct jar tests that mimic the anticipated operation and then to measure the sludge production directly.

8.3. SEPARATION OF RESIDUE

During treatment, the residue is usually removed in consecutive treatment steps, each producing residue in different quantities and concentrations.

8.3.1. Residue removal by filtration

Treatment plant operators know from experience that rapid sand filtration is optimal if the turbidity of the water to be filtered is kept between 2 and 5 NTU. Turbidity below this range signals the possibility of dosing too much coagulant; turbidity above may push up the required backwash frequency and loss of backwash water. So regardless of the raw water turbidity, the residue washed out from the filters remain almost constant – the spikes in raw water turbidity are buffered by the preceding sedimentation or dissolved air flotation steps.

The residue in filter washwater can be roughly approximated without resorting to direct measurement, instead using typical values encountered at South African treatment plants, shown in Example 8.2.

Example 8.2

A filtration plant treats a flow rate of 15 000 m^3/h with a total filtration area of 3 000 m^2. The filters are backwashed every 36 hours at a rate of 20 m/h for 10 minutes. Estimate the mass of residue removed after each filtration cycle, the residue concentration and the average residue production rate in the discharged washwater.

Volume of water filtered = 15 000 x 36 = 540 000 m^3 in 36 hours

Residue removed (4 NTU ≈ 4 g/m^3) = 540 000 x 4 = 2 160 000 g in 36 hours

Volume of backwash water = 20 x 3 000 x 10 / 60 = 10 000 m^3 in 36 hours

Residue concentration = 2 160 000 / 10 000 = 216 g/m^3

Residue production rate = 2 160 000 x 24 / 36 = 1 440 072 g/day =

526 tonne/year

At some treatment plants, operators have a "filter-to-waste" option. If the filtrate quality does not meet the required standard (commonly during rapid raw water quality changes or during plant commissioning), the filtrate is deflected away from the storage reservoir into the residue management system. This practice does not add any residue to the system but does add almost clean water that increases the hydraulic loading on the system.

8.3.2. Residue removal by sedimentation

The flocs entering the sedimentation tanks settle to form a layer of sludge on the bottom of the tank. For flat-bottomed tanks, the sludge is scraped to collection points or vacuumed from the floor by mechanical means. Some tanks with steeply sloped floors rely on gravity to guide the sludge to flow into troughs or hoppers from where it is pumped or drained out. Sludge is discharged by opening the discharge valves at regular intervals, each time allowing the sludge to wash out completely before the valves are closed. Alternatively, the valves automatically close after a preset time. At times of high sludge production, the frequency and time of desludging must be increased to prevent excessive build-up of sludge in the tank.

Although there are large differences in the details of different sludge removal systems, the solids concentration of the sludge should be roughly the same. A good rule of thumb for a well-operated treatment plant is to expect about 4% loss of raw water through the sludge from settling tanks, and another 4% through the washing of filters.

The average residue mass removed by sedimentation is obtained by subtracting the filtration residue from the total. In the absence of more specific information such as the desludging rates and intervals, the residue concentration can be estimated by using the typical loss rates provided in the previous paragraph, as shown in the example problem following.

Example 8.3

Estimate the residue concentration of the settling tank sludge for the same scenario used in Examples 8.1 and 8.2, with the added knowledge that the water loss from the settling tanks is 4.5% of the raw water flow.

Total residue in water = 30 960 kg/day (Example 8.1)

Residue removed by filtration = 1 440 kg/day (Example 8.2)

Residue removed by settling = 30 960 – 1 440 = 29 520 kg/day

Water loss through settling = 4.5% of 15 000 x 24 = 16 200 m^3/day

Residue concentration in sludge = 29520 x 1000 / 16 200 = 1 822 g/m^3

The results of Examples 8.1, 8.2 and 8.3 are summarised in Figure 8.1. The liquid flows of residue leaving settling and filtration are not so very different (two to three times), but that the mass flow from settling is vastly more than that from filtration (20 times). The concentration of solids in settled sludge is roughly ten times more in sludge than washwater.

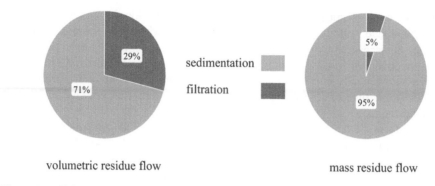

volumetric residue flow   mass residue flow

Figure 8.1   Relative volumetric and mass flows of residue from typical water treatment plants

Up to this point in the chapter, solids concentrations were expressed in mg/L or the equivalent g/m^3. This was convenient because it provided an easy link with units used for drinking water treatment and numbers are used that were convenient – neither too small nor too large for easy interpretation. For the rest of the chapter, however, we are dealing with concentrated residue with much higher concentrations. For residue management, the residue concentration is traditionally expressed as a percentage of dry solids on a mass/mass basis. One percent of dry solids is equal to 10 000 mg/L or g/m^3.

8.3.3. Residue removal by dissolved air flotation

It is more difficult to predict the concentration of the residue produced by dissolved air flotation. Dissolved air flotation captures the residue by small bubbles which rise to the surface, thereby collecting the residue in a concentrated float layer at the top of the reactor. There are two ways used to remove the float layer:

• By hydraulic flushing, which follows if the outlet of the tank is closed while the inlet is left open. This causes the tank level to rise to a point where water starts to spill over an overflow weir, dragging the float layer along. This method produces a relatively large volumetric flow of sludge with low concentration, although the concentration is still higher than the residue concentration produced by sedimentation.

• By mechanical removal, when a moving scraper travels over the surface of the tank, continuously or intermittently, and drives the float layer to a beach plate at the end of the tank, from where the residue enters the waste management system. This produces a much lower volume of sludge, which is sticky, heavy and more difficult to handle.

A summary of design values for these two options is reported in Table 8.1.

Table 8.1: Practical design values for different float layer removal methods[a]

	Hydraulic Flushing	Mechanical Removal
Concentration range	0.3% to 0.5%	1% to 3%
Maximum concentration to allow gravity flow	n/a	5%
Minimum overflow depth to keep weir clean	20 mm	n/a
Maximum flushing interval	10 hours	n/a
Scraper speed	n/a	60 m/h

[a] From Edzwald and Haarhoff (2012).

8.3.4. Variability of residue production

Up to this point in the chapter, the focus was on average residue production. It is evident that the main driver of residue production is the raw water turbidity, which is highly variable for South African surface water supplies. It follows that residue production will also be highly variable. If the plant records are available, the above calculations can be carried through for every day, week or month of the available data set, depending on how the data are available. From the results, it would be possible to construct a probability exceedance diagram.

When dealing with highly variable data sets, probability exceedance diagrams offer a useful method of analysis. They are not only useful for robust estimates of residue production, but for dealing with all data sets with high variability. The methodology is as follows:

- Compile all the data points into a linear sequence of values. The data points must preferably be taken at regular time intervals, for example daily, monthly or weekly.
- Use the PERCENTILE function in EXCEL to find the 10th percentile, 20th percentile, etc. all the way up to the 90th percentile. The minimum value is the 0th percentile and the maximum the 100th percentile. For better resolution, more intermediate percentile values can be calculated.
- Plot the percentile values to obtain the probability exceedance diagram.

Such an approach was taken for a South African water treatment plant known for large turbidity variations (Haarhoff et al., 2002). An initial probability exceedance diagram for residue production, based on the values of only two preceding years, indicated a 95th percentile of 1 300 g/m^3 and a 50th percentile of 100 g/m^3 at first – the values adopted for maximum and average values respectively. When the same exercise was repeated nine years later, then having more detailed data, the maximum and average values could be adjusted to 400 g/m^3 and 50 g/m^3 respectively. Both data sets indicated that the maximum was an order of magnitude more than the average.

8.4. COLLECTION OF RESIDUE

The residue streams from the different process units must be kept in suspension to prevent the solids from settling out as they flow to the next step. It is preferable to use open channels rather than pipes to be able to check on solids deposition and cleaning them out when necessary. Sharp corners and holding tanks must be avoided.

When particularly high solids concentrations are anticipated, for example for mechanical removal of the float layer scraped off dissolved air flotation tanks, the residue stream may become so thick that it will not flow under gravity, a limit reached at about 5% solids concentration. Such high percentages must be avoided by operational adjustments. If not possible, water jets must be provided to dilute the residue to the point where it flows freely.

8.5. CONCENTRATION OF RESIDUE

A residue concentration of sludge, typically 2%, is still very dilute. For instance, if the concentration can be increased to 10%, its volume will be reduced five times. Not only will the ultimate disposal volume be so much less, but 80% of the water can be recovered. There are two issues to consider under this heading – ways to thicken the residue, and concerns for the recycling of recovered water.

8.5.1. Residue thickening

At present, gravity thickening is the preferred method for thickening water treatment residue. The usual method in South Africa is to use sludge lagoons, where the residue streams from filtration and sedimentation are combined to flow into large shallow lagoons. The bulk of the residue settles to the bottom and relatively clean supernatant is either discharged into a natural watercourse or recycled back to the raw water inlet. After a year or more, the lagoon is taken offline and left for a few months in the dry season to dry out. When it reaches a consistency suitable for mechanical handling, it is excavated to leave the lagoon at its original level, ready for the next use cycle. The excavated material is treated or disposed of in a next process.

The design parameters, based on records over a period of nine years, of a series of sludge lagoons built in 1999 at the Vaalkop Water Treatment Plant are presented in Table 8.2 to illustrate a rational basis for their sizing and anticipated operation. These lagoons, under average conditions, should have a filling time of approximately 18 months.

Table 8.2: Design parameters for sludge lagoons at the Vaalkop Treatment Plant[a]

Average flow rate through treatment plant	150 000 m³/day
Average flow from sedimentation and filtration (9%)	13 500 m³/day
Residue flow (assumed 0.05 ton residue / ML)	7 500 kg/day
Residue concentration flowing into lagoons	0.556 kg/m³
Volume of one lagoon (one of three)	32 100 m³
Sludge holding capacity of lagoon (filled to 65% of depth)	20 865 m³
Residue concentration after settling in lagoon (assumption)	20%
Residue in lagoon when full	4 173 000 kg
Time to fill lagoon with sludge	556 days

[a] From Haarhoff et al. (2002)

The lagoons referenced in Table 8.2 are large with a total area of 2 hectare. If space does not allow their construction, an option remains to use a sludge thickener – a concrete or steel tank with a much smaller footprint designed to thicken the sludge quickly and effectively, similar to those utilised for mining and sewage sludges. These tanks have mechanical equipment such as radial scrapers or travelling bridges for the continuous removal of the thickened sludge. Under these conditions, it is necessary to assist the thickening process by adding a polymer to improve the settleability of the sludge.

Would the use of a polymer also improve the performance of conventional sludge lagoons? Such tests were performed before the design of the lagoons in Table 8.2. Anionic polymer was tested with sludge from settling tanks with a starting concentration of 3.9%. It took 20 hours to double the residue concentration without polymer, but only eight minutes with polymer. After 100 days, however, the concentration of both settled sludges, one treated (with polymer) and the other untreated sludge were both about 14%. For thickening, a polymer is necessary to allow appreciable thickening in a small tank with a short retention time. For sludge lagoons, where the sludge remains for months or even years, there seems to be little advantage in using polymer.

8.5.2. Recycling of recovered water

Water treatment plant managers obviously strive to recover spent filter backwash water. Not only does it save money by buying less raw water but also reduces return flows to public streams. At the same time, it is equally obvious that there should be water quality concerns when concentrated impurities removed from the water is recycled to the raw water inlet, thereby adding to the load on the treatment plant. Over the past two decades, these concerns were developed into detailed guidelines and regulations in some countries. In the USA, for example, the USEPA notified its intent to regulate filter backwash recovery in 1997 when it proposed an interim Filter Backwash Recycling Rule (FBRR) in 2000. It promulgated the FBRR in 2001 and finally followed it up with an FBRR Guidance Manual in 2002. Although filter backwash water is recovered at many South African water treatment plants, there is a lack of literature to explain the process or to gauge its performance, the exception being a summary of the performance of the filter backwash water recycling plant built by Rand Water at their Zuikerbosch Treatment Plant (Linde 2003).

An exhaustive study at 25 water treatment plants in the USA demonstrated that backwash water does more than just concentrate the residue. The concentrations of some dissolved contaminants are also elevated, summarised in Table 8.3.

Table 8.3: Deterioration in backwash water quality [a]

Parameter	Raw Water	Backwash Water
Dissolved organic carbon (mg/L)	2.4	8.0
Trihalomethanes THMs (µg/L)	0.6	55.0
Haloacetic acids HAA6 (µg/L)	1.9	46.1
Aluminium (mg/L)	0.72	14.7
Iron (mg/L)	1.2	8.7
Manganese (mg/L)	0.11	1.4
Zinc (mg/L)	0.03	0.1

[a] Averaged from USEPA FBBR Manual (2002).

The FBRR specifically calls for untreated backwash water to be returned to the head of the treatment plant to address the following concerns:

- The return of the recycle stream after the point of coagulant addition will dilute and alter the chemistry of the raw water stream, therefore affecting treatment efficiency.

- If the recycle is not subjected to coagulation and filtration, oocysts and other contaminants could pass through the filters.

To minimise flow disturbances, the recycle flow rate must be kept as even as possible and always be less than 10% of the plant flow. An alternative school of thought suggests that the recovered backwash water must rather be treated on its own to an appropriate standard, rather than inviting the risks involved with blending the recovered backwash water with the raw water flow. Rand Water took this approach at their Zuikerbosch Treatment Plant, where they opted to treat the backwash water to potable standard through flow equalisation, chemical dosing, lamella settling and rapid sand filtration (Linde, 2003).

8.6. TREATMENT OF RESIDUE

Although many options had been investigated for extracting some economic benefit from water treatment residue, real success stories remain to be reported. Two main approaches had been followed – extraction and reuse of the embedded coagulant; or use of the residue as an input to a subsequent manufacturing process.

An obvious target is to recover the coagulant trapped as precipitates within the residue, relevant for iron and aluminium-based coagulants. The strategy is to lower the pH to the point where the iron and aluminium precipitates are resolubilised, then getting rid of the remaining solids. This had been tried in Japan. By dropping the pH to pH 3 or less, between 60% and 80% of aluminium sulphate could be recovered, but the cost and potential contamination by the sulphuric acid consumed eventually stopped the practice. The recovery of ferric coagulants was even less successful – the pH had to be reduced to below pH 2 to attain 60% to 70% recovery (Doe, 1990). The recovery of lime from the precipitates formed by water softening requires a complicated scheme to resolubilise the

Mg^{2+} ions, separate the Ca^{2+} ions from the natural particles and recalcination to recover the lime as CaO. This process was never tried in South Africa.

An alternative recovery target is to use the clay particles contained in the residue. Numerous experiments were conducted with sludge from the Wiggins Treatment Plant of Umgeni Water to manufacture bricks and tiles (Boucher and Van Eeden 1994). Many difficulties were encountered, but eventually a method was developed to fire the brick with a special firing sequence and scrubbing the noxious gases generated. Although the study concluded that the process held promise for further refinement, it was not followed up.

8.7. DISPOSAL OF RESIDUE

Eventually, the residue must be finally disposed of. In earlier years, it was returned to the impoundments or streams where it came from, but environmental concerns have gradually put a stop to the practice. Nevertheless, there are no clear or simple alternatives yet. The most systematic approach to the problem in South Africa is a *Guideline for the Utilisation and Disposal of Water Treatment Residue* published by the Water Research Commission (Herselman et al., 2013). Although the emphasis of this document is largely focused on the application of the sludge onto agricultural land, it provides a useful framework for developing a strategy for dealing with water treatment residue. It also provides an overview of the legal framework for residue disposal in South Africa.

The *Guideline* identifies five disposal options:

- Land application, either on agricultural land, in forests, or for land reclamation;
- On-site disposal (ultimate disposal within the boundaries of treatment plant);
- Off-site disposal (ultimate disposal outside the boundaries of treatment plant);
- Discharge to wastewater treatment systems; and
- Reuse of the residue (coagulant recovery, bricks, cement, etc).

The direct discharge of the residue to the source stream is not considered an environmentally responsible option anymore and not provided for in the *Guideline*. The requirements for the second option, namely that of on-site disposal, are briefly summarised here. The requirements for the on-site disposal of the residue are grouped in three categories – those pertaining to the properties of the residue; those pertaining to the selection and properties of the disposal site; and those pertaining to the ongoing monitoring of the disposal site.

The initial characterisation of the residue calls for the measurement of eight metals (arsenic, cadmium, chromium, copper, lead, mercury, nickel, zinc) to check that the residue complies with the "Pollutant Class A" classification for wastewater sludge. In addition, should the raw water have an elevated uranium concentration, the uranium concentration in the sludge should also be tested. Should all the metals register below the prescribed limits, then on-site disposal of the residue is permissible, subject to further analyses discussed in the next paragraph. Should one or more metals register above the limits, then on-site disposal is still permissible, but subject to maximum load restrictions.

Once the residue passed the initial metals tests above, more detailed tests are required to determine the leachable fraction of metals and trace elements in the residue. The main test is the Toxicity Characteristic Leaching Procedure (TCLP) which is used to calculate the

Estimated Environmental Concentration (EEC). If the EEC is less than the Acceptable Risk Level (ARL), then the residue can be delisted (an administrative procedure) and disposal is permissible. The TCLP, EEC and ARL are applied to each of the trace metals. If any one of these metals do not meet the EEC < ARL test, then the residue must be treated with lime and the test repeated. A problem with elevated manganese concentration, which impacts on the classification of the residue, had previously been encountered in South Africa.

There are technical guidelines for the disposal site:

- Topography should be suitable (to prevent erosion, ponding and runoff).
- Soil properties should be suitable (less than 20% clay would require the lining of the site, pH should be above pH 6.5 to prevent the mobilisation of metals).
- Sensitive areas should be avoided (sinkholes, fault zones, seismic zones).
- Depth to aquifer should preferably be more than 5 m.
- Disposal site to nearest surface water or borehole should be more than 200 m.

Once a disposal site is operational, the following monitoring programme is suggested:

- Quarterly TCLP tests on the residue until it is shown that its quality is constant; thereafter, the frequency can be relaxed.
- Quarterly tests on the groundwater upstream and downstream of the site. The frequency can be relaxed when the water table is deeper than 5 m and the clay content of the soil 35%.
- Surface water monitoring 2050 m upstream and downstream of the site during rainy periods, unless the runoff is intercepted or diverted by cut-off drains.
- Every two years, samples of the underlying soil should be tested for metals.

The *Guideline* is useful in that it provides a systematic checklist for safe and responsible residue disposal. The important issues are addressed. The *Guideline*, however, is very strict and very few, if any, of the water treatment plants in South Africa comply with its requirements. The monitoring requirements are stringent and expensive and certainly not complied to at almost all South African water treatment plants.

8.8. FUTURE CHALLENGES

8.8.1. More stringent monitoring

Residue management is a neglected aspect of water treatment practice. Practical systems mostly evolved from empirical rules and expediency. Quantitative surveys and plant comparisons of residue management systems were rarely made and almost never reported. This will not be allowed to go unchecked anymore, as environmental controls and legislation are steadily getting more stringent. Rational analysis and perhaps radical thinking are now required as a matter of urgency. Holistic design, incorporating both treatment efficiency and residue management, is required.

8.8.2. Liquid residue from membrane filtration

Membrane treatment has grown logarithmically since the 1980s. The number of membrane treatment plants was about 100 in 1980, 1 000 in 1990 and 10 000 in 2000 (Pennsylvania Department of Environmental Protection N.d.). Membranes come with different pore sizes, with microfiltration (MF) the coarsest, through ultrafiltration (UF) and nanofiltration (NF) to reverse osmosis (RO), which is the finest. As the pore size decreases, more energy is required to push the water through the membrane. The water recovery (water volume passing through the membrane in relation to the water volume applied to the membrane) also depends on the pore size with 95% to 98% for MF, 90% to 95% for UF, and 75% to 90% for NF. For RO, the water recovery is also dependent on the salinity of the water. For freshwater, the water recovery is in the 50% to 70% range and even lower for seawater. The transmembrane pressure and water recovery are two major constraints for using RO more widely and receiving much research and development, so the values above are used for illustration only.

Membrane fouling is inevitable. At frequent intervals (depending on the membrane type, the raw water quality and the pre-treatment), the membranes must be cleaned in place (CIP). Depending on the type of fouling, CIP is done with strong chemical solutions. For mineral scaling, citric acid is typically used; for organic fouling sodium hydroxide; and for microbiological fouling a biocide.

From the perspective of water treatment residue, it is evident that membrane treatment plants produce two types of liquid residue. The first is the water rejected during normal operation. If water which is mildly brackish (at say TDS of 500 mg/L) is filtered by RO and 70% is recovered, a waste stream of 30% of the raw water flow is generated with TDS of 500x100/30 = 1 667 mg/L. The second type of residue is generated during CIP. Here a much smaller waste stream is generated of contaminated concentrated industrial chemicals. In both these cases, it is fairly easy to predict the volumes and composition of the waste stream with the same approach used in the rest of this chapter.

However, there are no easy options for dealing with these waste streams. At the moment there are no proven options for the inland cities of South Africa. Where would the polluted waste streams go if Rand Water, for example, would switch completely to membrane filtration? Coastal cities do have the option of discharge to the ocean, but in the Persian Gulf, where RO is widely used at large scale, voices are already rising against these discharges into the ocean for their environmental impacts (Gies, 2019).

8.9. REFERENCES

Boucher, P.S., and Van Eeden, J.J. 1994. *Investigation of Inorganic Materials Derived from Water Purification Processes for Ceramic Applications*. Report 538-1-95. Pretoria: Water Research Commission. http://wrcwebsite.azurewebsites.net/wp-content/uploads/mdocs/538-1-95.pdf

Doe, P.W. 1990. "Water Treatment Plant Waste Management". In *Water Quality and Treatment*, 4th Edition. Denver: American Water Works Association.

Gies, E. 2019. "Slaking the World's thirst with Seawater Dumps Toxic Brine in Oceans." *Scientific American* 7 February. https://www.scientificamerican.com/article/slaking-the-worlds-thirst-with-seawater-dumps-toxic-brine-in-oceans/

Haarhoff, J., Van Heerden, P., and Van der Walt, J.J. 2001 "Sludge and Washwater Management Strategies for the Vaalkop Water Treatment Plant." *Water Science and Technology*, 44(6): 73–80.

Herselman, J.E. 2013. *Guidelines for the Utilisation and Disposal of Water Treatment Residues.* Report TT559/13. Pretoria: Water Research Commission. http://www.wrc.org.za/wp-content/uploads/mdocs/TT%20559-13.pdf

Linde, J.J. 2003. *Evaluation of a Filter Backwash Recovery Plant to Establish Guidelines for Design and Future Operation*, WRC Report 920/1/03. Pretoria: Water Research Commission. http://www.wrc.org.za/wp-content/uploads/mdocs/920-1-03.pdf

Pennsylvania Department of Environmental Protection. N.d. "Drinking Water Operator Certification Training Module 19: Membrane Filtration". http://files.dep.state.pa.us/Water/BSDW/OperatorCertification/TrainingModules/dw-19_membrane_wb_10_07.pdf).

United States Environmental Protection Agency. 2002. Filter Backwash Recycling Rule Technical Guidance Manual. https://nepis.epa.gov/Exe/ZyPDF.cgi?Dockey=200025V5.txt.

PART III

RECREATIONAL WATER

9

QUALITY

9.1. INTRODUCTION

Municipalities all over the world provide public areas where the inhabitants play and relax in different ways. Wherever possible, city planners endeavour to locate these public areas at or near open water bodies because water is a natural attraction. Beaches, parks, sporting areas, picnic spots, walkways and paths, city squares and angling areas are all associated with oceans, lakes, riverbanks and fountains. Municipalities are obliged to build, maintain and manage these public areas, including the water bodies.

This chapter demonstrates that it takes a scientific approach and good technical understanding to interpret and apply the regulations governing public safety and specifically the potential water quality issues that may emerge (see United States Environmental Protection Agency 2012 as an example of a comprehensive framework dealing with this topic). It also sets the stage for Chapter 10 on the management of swimming pools to avoid the health risks highlighted here.

9.2. GENERAL SAFETY CONCERNS

Exposure to water poses more dangers than just those associated with poor water quality. The most extreme danger faced by bathers is drowning, a major cause of death worldwide, particularly for male children. Drowning may be associated with swimming as well as with intermittent water contact, such as recreational use of watercraft (yachts, boats, canoes) and fishing. Alcohol consumption is one of the most frequently reported factors contributing to the drowning of adults whereas lapses in parental supervision are most frequently cited for children.

Most sports-related spinal cord injuries are associated with diving. Injuries from diving incidents are almost exclusively located in the cervical vertebrae, resulting in quadriplegia or paraplegia. Alcohol consumption contributes significantly to the frequency of injury. Other bodily injuries associated with recreational water use activities include brain and head injuries, fractures, dislocations, cuts and lesions.

Outdoor public areas expose visitors to solar radiation. Overexposure to the sun may result in acute and chronic damage to the skin, eyes and the immune system. The most noticeable and familiar effect of excessive solar exposure is inflammation of the skin, commonly termed sunburn. Chronic effects include two major public health problems, namely, skin cancers and cataracts. Chronic exposure further leads to degenerative changes in the skin (freckles, for example) and accelerated skin ageing.

Sudden immersion in cold water, other than the immediate shock familiar to all, could lead to cold-water immersion death. The most susceptible parts of the population are the very young, the elderly, patients using drugs that interfere with temperature regulation, people suffering from pre-existing chronic diseases and frequent consumers of alcohol.

9.3. EXPOSURE TO WATER AND ASSOCIATED RISKS

9.3.1. Degree of exposure

The users of public spaces are exposed to water in different degrees. Three broad categories of exposure are recognised (Department of Water Affairs and Forestry 1996):

- Full contact recreation (swimming) involves full-body water contact caused by swimming and diving, with a high probability of ingesting water directly. Children are the most vulnerable to infectious diseases. Children, even if they are not completely healthy, are still inclined to swim, thereby infecting others.

- Intermediate-contact recreation encompasses all forms of contact recreation excluding activities described for full contact recreation. It is a broad class and includes activities which involve a high degree of water contact, such as water skiing, canoeing and angling and those which involve relatively little water contact, such as paddling and wading. Angling is a common and popular recreational use of inland waters, often involving direct exposure to water as well as indirect exposure through the handling of fishing gear and fish. The range of activities requires that some discretion must be used in applying the guideline. A more stringent approach is necessary where water contact is frequent and relatively extensive whereas a less stringent approach can be adopted if water contact is infrequent and minimal. The intermediate contact category is less susceptible to poor water quality because full immersion is less frequent. On the other hand, strenuous full contact water sports are generally practised by water users in a good state of health.

- Non-contact recreation encompasses all forms of recreation which do not involve direct contact with water. It includes activities such as picnicking and hiking on the shores of a water body. These activities concern themselves predominantly with the scenic and aesthetic appreciation of water. The economic value of recreational water bodies is often closely related to scenic appreciation since this is a major factor in determining the value of waterfront properties. Since no water contact occurs, public health effects associated with water contact are of little relevance.

9.3.2. Turbidity

Turbidity is caused by suspended solids derived from silt or organic debris that could reduce visibility within the water body to such an extent that underwater hazards may not be visible, creating dangerous situations for swimmers and divers as well as damage to equipment such as boats and skis. Some rivers and impoundments have permanent low visibility owing to the nature of their catchments, or because of the source of their feed water and turbulence in the impoundment. In some cases, the reduction in visibility is only temporary resulting from flash floods carrying high silt loads.

Little can be done to mitigate poor visibility in naturally turbid waters, except to post warning signs to warn against accidents. For waters with usually good visibility, but subject to occasional peaks in turbidity owing to flash floods or upstream discharges, diversion weirs could be used to bypass turbid water past specific recreational sites until the turbid slug of water has passed.

9.3.3. Algae

In the case of benthic algal growths (meaning algae that grow on surfaces), especially those secreting gelatinous material, submerged surfaces such as rocks, concrete or wood may become very slippery, posing a threat to human safety. Participants in full contact recreational activities who have suffered minor mishaps as a result, usually recover fully. Blooms of planktonic algae (meaning algae that are suspended and not attached to surfaces) may reduce the visibility to the point where they pose the same dangers as those outlined for turbidity.

The effects of both benthic and planktonic algae may be mitigated to a certain extent by mechanical (for benthic algae) or biocidal treatment (for both planktonic and benthic algae) programmes. Such biocidal treatment programmes may not be effective in flowing waters. Caution, however, should be exercised when biocides are used as these may give rise to skin irritations or to the release of toxins from the dying organisms.

9.3.4. Nuisance plants

The profuse growth of macrophytes in impoundments or along riverbanks, while possibly aesthetically pleasing, may provide hazardous situations for participants of full contact recreational activities because of snagging and entanglement. Therefore, the application of herbicides to eradicate macrophytic plant growth in impoundments or along stream margins will eliminate or ameliorate hazardous situations for full contact recreational water users. However, the use of herbicides may cause skin irritations or the release of toxins from the decaying plants, as mentioned before.

Besides the risk of entanglement, a more serious concern are the nuisance plants that render water bodies aesthetically displeasing or give rise to discomfort for full contact recreational water users. Such plants may give rise to unsightly or odorous substances, and if present in large numbers, may constitute a hazard to human health and safety. Heavy blooms of *Microcystis aeruginosa* (a species of cyanobacteria or blue-green algae common in South Africa) release phytotoxins when they die, which pose a health hazard. Decaying algae also give rise to foul odours and unsightly masses of decaying vegetation when they form a thick crust covering the water surface.

The effects of nuisance plants are usually seasonal, often persisting for a long time after their growth period. The effects of nuisance plants may be mitigated by either exploiting environmental factors to inhibit growth, or by mechanical, chemical or biological control programmes to reduce or eliminate the presence of these plants.

9.3.5. Floating matter

The presence of shoreline litter and floating matter of human and natural origin detracts from the aesthetic enjoyment of water bodies. Submerged refuse also presents a danger to full contact recreational water users. Floating matter consists of waste oil and grease, plastic containers and bags, bottles, cans, metal containers and domestic refuse. Some objectionable floating matter may also be generated naturally through decaying vegetation. Although difficult to achieve in some instances, education in environmental awareness may decrease the dumping of litter in or near water bodies. Organised campaigns to clean up the environment, both terrestrial and aquatic, are often used to reduce the amount of objectionable floating matter.

9.4. HUMAN HEALTH CONCERNS

9.4.1. Skin, ear and eye irritation

Although it may not be a serious or life-threatening concern, it is appropriate to highlight the effect of pH on the eyes of humans. We cannot do much about the pH of large natural water bodies, but pH is one of the key parameters for proper swimming pool management which is covered in the next chapter. Table 9.1 provides a convenient summary:

Table 9.1: pH guidelines for full contact recreational water [a]

pH Range	Effects
less than 5.0	Severe eye irritation. Skin, ear and mucous membrane irritation likely. Adverse aesthetic taste effects expected if water accidentally swallowed.
5.0 – 6.5	Swimming generally acceptable. Some eye irritation. Skin, ear and mucous membrane irritation unlikely.
6.5 – 8.5	Minimal eye irritation. Within the buffering capacity of the lachrymal fluid of the human eye. Skin, ear and mucous membrane irritation absent.
8.5 – 9.0	Swimming acceptable. Some eye irritation. Skin, ear and mucous membrane irritation possible. Adverse aesthetic taste if water swallowed accidentally.
more than 9.0	Eye irritation increasingly severe, skin, ear and mucous membrane irritation. Adverse aesthetic taste expected, if water swallowed accidentally.

[a] Department of Water Affairs and Forestry 1996

Through contact with the skin or penetration of the ear, microbially or chemically contaminated water may cause skin infections, ear infections and irritations. Such infections may be chronic or acute, depending on the nature and source of the contamination or the state of the person's immune system. The use of ear plugs by participants in full contact recreational activities may prevent or reduce infections of the outer and middle ear.

9.4.2. Waterborne diseases

The water body used for full contact recreational activities, if microbially contaminated, is a potential source of infectious diseases. Such diseases are contracted either by ingestion or through contact with the skin, especially mucous membranes. Depending on the type of waterborne disease and on the physical health of the person infected through full contact recreational activities, the person may either recover completely from the disease, or suffer permanent harm or damage from the disease, or if severe enough, the person may die. Waterborne gastroenteric diseases are contracted from the ingestion of water contaminated with pathogenic faecal organisms, polluted by algal toxins or other chemical pollutants. Full contact users suffering from gastroenteric disorders usually recover fully from the effects following treatment.

Remedial measures, such as removing or controlling the source of contamination, should either eliminate or mitigate the effects of infectious diseases. Banning any form of full contact recreation in a contaminated water body is the best way to eliminate the risk of contracting waterborne diseases from the affected water body.

9.4.3. Using the HACCP approach

The Hazard Analysis and Critical Control Points (HACCP) concept was developed in the food industry, subsequently applied to potable water treatment and distribution. This was explained in some detail in Chapter 2, where it was shown that any HACCP programme must address seven steps. The World Health Organisation (WHO) offered an interpretation of the seven steps in relation to recreational waters, shown in Table 9.2.

Table 9.2   HACCP applied to recreational water [a]

Core Principles	Interpretation for Recreational Water
Hazard analysis	Identify different types of faecal pollution and potential points of entry, their exposure risks and preventative measures
Critical control points	Identify points at which management actions can be applied, such as sewage discharge points, treatment plants and illegal connections
Critical limits	Determine measurable control parameters and their critical limits
Monitoring	Establish a monitoring programme to detect exceedance beyond limits
Management actions	Develop actions to reduce or prevent exposure should limits be exceeded.
Validation / verification	Obtain evidence that management actions will improve water quality and limit exposure from the literature and own audit data
Record keeping	Retain monitoring records in a format that permits external audit and compilation of annual statistics

[a] Summarised from World Health Organisation 2003

9.5. MICROBIOLOGICAL ASSESSMENT

9.5.1. Sewage contamination

From a microbiological perspective, the most pressing concern regarding recreational waters is the potential contamination by raw or poorly treated sewage. There are three possible pathways to link sewage with recreational water:

- Direct discharge of raw or treated sewage into the water body.
- Raw or treated sewage discharged into a tributary river feeding the water body. Identification of these discharges requires thorough knowledge and monitoring of the catchment area.
- Bathers, especially smaller children, may introduce faecal matter directly into the water body.

Raw sewage is a host to numerous pathogens. Table 9.3 provides a list of the most important pathogens, indicating their typical concentrations in raw sewage.

Table 9.3: Pathogens in raw sewage [a]

Pathogen / Index Organism	Disease / Role	Numbers / 100 mL
Bacteria		
Campylobacter spp.	Gastroenteritis	10^4–10^5
Clostridium perfringens spores	Index organism	6×10^4 – 8×10^4
E.coli	Index organism	10^6–10^7
Faecal streptococci	Index organism	4.7×10^3–4×10^5
Salmonella spp.	Gastroenteritis	0.2–8 000
Shigella spp.	Bacillary dysentery	0.2–1 000
Viruses		
Polioviruses	Index organism / poliomyelitus	180–500 000
Rotaviruses	Diarrhoea, vomiting	400–85 000
Parasitic protozoa		
Cryptosporidium parvum oocysts	Diarrhoea	0.1–39
Entamoeba histolytica	Amoebic dysentery	0.4
Giardia lamblia cysts	Diarrhoea	12.5–20 000
Helminths (ova)		
Ascaris spp.	Ascariasis	0.5–11
Ancylostoma spp. and *Necator* sp.	Anaemia	0.6–19
Trichuris spp.	Diarrhoea	1–4

[a] From World Health Organisation 2003

9.5.2. Indicator organisms

It would be impractical to monitor all possible pathogens, both technically and economically. Instead, indicator organisms are used for routine monitoring. The ideal indicator organism should (South African Guideline 1996):

- be suitable for all types of water;
- be present in sewage and polluted waters whenever pathogens are present;
- be present in numbers that correlate with the degree of pollution;
- be present in numbers higher than those of pathogens;
- not multiply in the aquatic environment;
- survive in the environment longer than the pathogens;
- be absent from unpolluted water;
- be detectable by practical and reliable methods; and
- not be pathogenic and be safe to work with in the laboratory.

There is no single indicator organism that could mimic the wide variety of pathogens in raw sewage. Multiple indicator organisms are therefore used, each with its own advantages and disadvantages. A good monitoring programme would include a selection of indicator organisms to get a better description of the water quality. The South African guidelines suggest that a selection should be made from the following indicator organisms:

- Total coliform bacteria: Indicator of the general sanitary quality of water since this group includes bacteria of faecal origin. However, many of the bacteria in this group may originate from growth in the aquatic environment. Used to evaluate the general sanitary quality of drinking water and related waters, for example swimming pool water, also routinely used for the monitoring of drinking water supplies.

- Faecal coliform bacteria: Indicator of probable faecal pollution of water since this group is more closely associated with faecal pollution than the broader total coliform group. Some faecal coliforms may not be of faecal origin. It is used to evaluate the quality of wastewater effluents, river water, seawater at bathing beaches, raw water for drinking water supply, recreational waters as well as water used for irrigation, livestock watering and aquaculture.

- *E.coli*: Highly specific indicator of faecal pollution which originates from humans and warm-blooded animals.

- Enterococci (faecal streptococci): These bacteria always appear in human and animal faeces, but in lower numbers than total or faecal coliforms and are more resistant than coliform bacteria. Not all faecal streptococci are of faecal origin, resulting in taxonomic regrouping of this group in recent years. Enterococci comprise a subgroup of faecal streptococci and include predominantly faecal streptococci of proven faecal origin. It is used in evaluation of treatment processes and recreational waters

- Bacteriophages: These phages occur in large numbers in sewage, and are detectable by relatively simple, economic and rapid techniques. Their numbers may increase in certain water environments suitable for the growth of host bacteria. Somatic coliphages indicate faecal pollution, and their incidence and survival in water environments would seem to more closely resemble that of human viruses than faecal bacteria. Survival and incidence of bacterial viruses (phages) in water environments resemble that of human viruses more closely than most other indicators commonly used.

9.5.3. Suggested indicator limits

The South African guideline provides guidance on all the parameters that are implicated in recreational water quality. Many of the guidelines are qualitative where analytical measurement is difficult or impossible, such as odour, floating objects and nuisance plants. For microbiological parameters, some quantitative guidelines are provided for full contact recreation (Table 9.4) and intermediate contact recreation (Table 9.5). For non-contact recreation, the microbiological parameters do not apply and the main concern is about the aesthetic parameters, where qualitative assessment would be sufficient.

Table 9.4: Risk indicators for full contact recreation [a]

	Little Risk	Slight Risk	Some Risk	Risk
Enteric viruses (TCID$_{50}$/mL)	0	1–10		>10
Coliphages (#/100mL)	0–20	20–100		>100
Faecal streptococci (#/100mL)	0–30	30–60	60–100	>100
E.coli (l/c c) (#/100mL)	0–130	130–200	200–400	>400
Faecal coliforms (#/100mL)	0–130	130–600	600–2000	>2000

[a] Adapted from Department of Water Affairs and Forestry 1996

Table 9.5: Risk indicators for intermediate contact recreation [a]

	Little Risk	Some Risk	Risk
Faecal streptococci (#/100mL)	0–230	230–700	>700
Faecal coliforms (#/100mL)	0–1000	1000–4000	>4000

[a] Adapted from Department of Water Affairs and Forestry 1996

9.6. BLUE FLAG BEACHES

9.6.1. The Blue Flag programme

The international Blue Flag programme is one of the international success stories of integrated environmental management. The programme started in 1985 in France, was soon embraced within the European Union and in 2001 South Africa became the first country outside Europe to join the Blue Flag programme. There are three main Blue Flag categories, which apply to beaches, marinas and tourism boats respectively. In February 2018, 4 423 Blue Flag sites were recorded in 45 countries (Blue Flag 2020). South African currently has 44 Blue Flag beaches.

The Blue Flag "year" in the southern hemisphere runs from 1 November to 31 October. In addition, each site is only registered for its Blue Flag "season", which could run from one or two months (say December and January) to the full year. Sites are required to reapply each year to be considered for the Blue Flag award and applications are reviewed by national and international juries.

The Blue Flag programme offers many benefits – improved tourism facilities, enhanced management of coastal ecosystems, increased awareness of the coast and capacity building of coastal municipalities. The trusted eco-label provides local beachgoers, domestic and international holidaymakers with the assurance of world class beaches offering safe, clean and well-managed facilities.

9.6.2. Blue Flag criteria

The award of Blue Flag status follows a rigorous process of having to demonstrate that every single one of 33 different criteria are met. Most of these criteria are mandatory – if any of these criteria are violated, Blue Flag status is immediately and automatically withdrawn for the rest of the year. Some criteria are put forward as "guidelines", but any violation is noted and if more than one of these guidelines are violated, cancellation of Blue Flag status could follow.

The 33 guidelines are grouped under four main categories:

- Environmental Education and Information (6 criteria)
- Water Quality (5 criteria)
- Environmental Management (15 criteria)
- Safety and Services (7 criteria)

Five of the 33 guidelines pertain specifically to water quality and are discussed in more detail.

Criterion 7

The beach must fully comply with the water quality sampling and frequency requirements:

There should be at least one sampling point for the beach, where the concentration of bathers is anticipated to be the highest. For each sampling point, there must be no more than 31 days between any two water samples during the Blue Flag season. A minimum of five samples must be taken, evenly distributed during the season. The first sample must be taken within 31 days before the official starting date of the Blue Flag season. In the event of short-term pollution, one additional sample is to be taken to confirm that the incident has ended.

Criterion 8

The beach must fully comply with the standards and requirements for water quality analysis:

An independent person, officially authorised and trained for the task, must collect the samples. An independent laboratory must carry out the analysis of the bathing water samples. The laboratory must be nationally or internationally accredited to work according to European (FEN) or International (ISO) standards. Water quality results must be provided to the National Operator (for South Africa, WESSA – the Wildlife and Environment Society of South Africa) as soon as they are made available but not later than one month after the sample has been taken. The water quality results for the previous four seasons must accompany all applications. In order to be eligible for the Blue Flag, the beach must show that the bathing water quality standards were met during the previous seasons.

Criterion 9

Industrial, wastewater or sewage-related discharges must not affect the beach area:

A bathing water profile must be compiled for every Blue Flag beach. A bathing water profile includes identification of potential sources of pollution, a description of the physical, geographical and hydrological characteristics of the bathing water, and an assessment of the potential for cyanobacteria and algae formation. It is recommended that there should not be any industrial, urban wastewater or sewage-related discharges into the Blue Flag area or immediate surrounding area. If there are discharge points in the designated beach area, these must be documented at the time of application. The collection, treatment and discharge of urban wastewater in the community must meet national/ international standards and comply with national/international legislation.

Criterion 10

The beach must comply with the Blue Flag requirements for the microbiological parameter *E.coli* and streptococci:

For the evaluation of an applicant beach, the Blue Flag Programme requires 95th percentile compliance with the limit values presented in Table 9.6. This is in accordance with the EU Bathing Water Directive (2006) as well as the recommendation of WHO. The percentile must be calculated for each parameter and met for each parameter. For example, if the 95th percentile is below the limit values for *E.coli* but not for faecal streptococci then the beach cannot be awarded with the Blue Flag.

Table 9.6 Blue Flag microbiological limits [a]

	Coastal Waters	Inland Waters
Faecal streptococci (#/100mL)	<100	<200
E.coli (#/100mL)	<250	<500

[a] Wildlife and Environment Society of South Africa 2019

Criterion 11

The beach must comply with the Blue Flag requirements for physical parameters:

There must be no oil film visible on the surface of the water and no odour detected. Ashore and on land, the beach must be monitored for oil and emergency plans should include the required action to take in case of such pollution. No floatables may be present, such as tarry residues, wood, plastic articles, bottles, containers, glass or any other substances. Immediate action should be taken if abnormal changes are detected. Should physical and chemical pollution be detected repeatedly, the Blue Flag must be taken down for the remainder of the season and the beach will not be eligible for the Blue Flag the following year.

9.6.3. Future development of the Blue Flag programme

One of the obstacles to the Blue Flag programme in South Africa is the significant difference between the east and west coasts of the country. Few countries in the world have sub-tropical waters flowing on the one side and polar water flowing on the other, an issue that WESSA has brought to light.

In South Africa, currently nine beaches in KwaZulu-Natal (KZN) have Blue Flag status when compared to 28 in the Western Cape. It is more difficult to comply with the water quality standards in beaches with warmer water than in colder water. The implication of tropical water versus cold water will become more of an issue as more countries from the southern hemisphere join the Blue Flag programme.

Another challenge arises from the duration of the Blue Flag season. KwaZulu-Natal, with its warmer climate and water, has a year-round tourism season of 12 months, while the Western Cape with its short tourism period has a season of only four months. It is therefore more difficult to sustain the high Blue Flag standards in KwaZulu-Natal than in the Western Cape.

Some municipalities have found the Blue Flag beach programme to be a double-edged sword. Beaches in Durban and Margate, for example, received unwanted publicity after losing their Blue Flag beach status. The WESSA spokesperson nevertheless encouraged participation in the Blue Flag programme:

> It takes courage to be part of the Programme, if something happens and your beaches lose Blue Flag status, everyone will know about it, even though there might be lots of factors beyond your control.

9.7. REFERENCES

Blue Flag. 2020. *Blue Flag Beach Criteria and Explanatory Notes 2020.* www.blueflag.global.

World Health Organisation. 2003. *Guidelines for Safe Recreational Water Environments Volume 1: Coastal and Fresh Waters.* Geneva: World Health Organisation. http://www.who.int/water_sanitation_health/bathing/srwg1.pdf

United States Environmental Protection Agency. 2012. *Recreational Water Quality Criteria.* Report 820-F-12-05. https://www.epa.gov/sites/production/files/2015-10/documents/rwqc2012.pdf

Department of Water Affairs and Forestry. 1996. *South African Water Quality Guidelines: Volume 2 – Recreational Water Use.* Pretoria: Department of Water Affairs and Forestry. http://www.dwa.gov.za/Documents/Other/RMP/RWUM/RWU_GP6.pdf

SWIMMING POOL MANAGEMENT

10.1. INTRODUCTION

Most towns and cities in South Africa have municipal swimming pools, mostly with dedicated staff to take care of their daily operation. From time-to-time, however, they are subject to problems of some sort. In cases of mechanical or power failures, the remedial action is straight-forward – just find and fix or replace the faulty component. But often, pools develop water quality problems which are not a result of mechanical failure but due to other reasons. Troubleshooting of swimming pool problems requires a basic understanding of both pool hydraulics and water chemistry.

This chapter firstly provides an overview of the main swimming pool components and their hydraulic parameters. Secondly, special attention is directed towards water quality problems and the wide range of pool chemicals available to cure or alleviate the problems – what they are, how they work and when they could offer a solution to the problem.

10.2. SWIMMING POOL DESIGN

10.2.1. Schematic layout

Figure 10.1 shows the schematic layout of a typical domestic swimming pool. Except for the pool itself, there is a sand filter and a recirculation pump. The recirculation pump draws water from the pool outlet box, pumps it into the top of the filter to force it down through a sand layer and back to the pool through one or more side inlets. This is the normal circulation mode. When the filter becomes dirty, the outlet of the recirculation pump is switched to the bottom of the filter, reversing the flow through the sand bed to flush out the impurities trapped during the circulation mode. This is the backwash mode. The circulation mode normally continues for one to seven days (depending on climate and pool usage) while the backwash period has a duration of 5 to 15 minutes.

For domestic pools, there are no rigorous design criteria. It is up to the pool contractor to match the size of the filter and the capacity of the recirculation pump with the size of the pool. The pool, filter and pump are offered as a single commercial package to the customer. Although the rest of the chapter focuses on larger, municipal pools which should be subject to much closer specification and stricter regulation, all the principles and possible solutions apply to domestic pools as well — the chemistry stays the same.

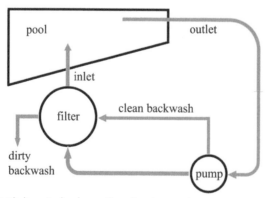

Figure 10.1 Schematic layout of a domestic swimming pool

Municipal pools are slightly more sophisticated than domestic pools, although their operation is the same:

- They are much larger, so more care must be taken to avoid short-circuiting, which happens when the filtered inlet water runs straight to the outlet, leaving large parts of the pool as stagnant areas. In addition, the flow rates are much higher; so, more care must be taken to avoid high water velocities at inlets and outlets.

- There are many more bathers in the water, for hours on end. In technical language, their bathing load is much higher, therefore also much more pollution of the water.

- As it is a public facility, it is regulated more rigorously to absolve the owner from liability claims in the case of accidents.

A schematic layout of a municipal pool is shown in Figure 10.2. The only difference from Figure 10.1 is that there are two outlets from the pool and that a balancing tank is provided between the pool and the pump. The bulk of the pollution by bathers and the atmosphere is carried in the top 75 mm of the water in the pool. Therefore, it makes sense to decant the surface layer off the pool as quickly as possible. The remaining water is taken from the pool through one or more bottom outlets on the floor of the pool. The surface and bottom outlets drain towards a common balancing tank, from where the water is pumped through the filters.

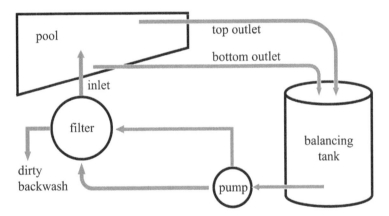

Figure 10.2 Schematic layout of a municipal swimming pool

10.2.2. Pool volume

The pool volume is the master variable for pool design – all the other elements are sized accordingly. The volume is determined by either calculation, or direct measurement:

- If the pool has a clearly defined geometry (meaning that you can make a simple dimensioned drawing of the pool shape), the volume is calculated by splitting the pool into rectangular and triangular prisms, calculating the volume of each and adding up.

- Domestic pools often have more artistic shapes; for example, kidney-shaped pools with rounded edges. To enable automatic pool cleaners to reach both the sides and bottom, domestic pools have a rounded, gradual transition between the sides and the bottom. For these cases, *in situ* measurement might be easier. Overlay a rectangular grid of lines over the pools (a hosepipe might even do), with about 500 mm between adjacent lines. Then measure the water depth at every node of the grid and average all the measured depths to get the average depth. By counting the number of squares in the grid, the area is determined next. Calculate the pool volume by multiplying the average depth with area.

The largest pools (Olympic-sized swimming pools are 55 m long and 25 m wide) hold about 2 500 m³, while the water volume of domestic pools range from about 30 to 70 m³.

Example 10.1

A grid of 600 mm by 600 mm is overlain on a small plunge pool. The water area includes 13 full squares, 18 partial squares and 19 nodes of the grid. The average depth of the 19 depth measurements is 1.39 m. Calculate the pool volume.

Area of one full square = 0.6 x 0.6 = 0.36 m²

Pool area therefore = (13 + 18/2) x 0.36 = 7.92 m²

Average depth = 1.39 m

Pool volume = 1.39 x 7.92 = 11.0 m³ or 11 000 L

10.2.3. Circulation rate

The pool water is kept clean by circulating it continuously through the filter. The more often the water passes through the filter, the faster the impurities are removed from the pool. This is ensured by specification of a maximum turnover time. The maximum turnover time depends on the circulation rate. The relationship is provided by:

$$Q = \frac{V}{T}$$	Equation 10.1
Q = circulation rate, expressed in m³/h V = pool volume, expressed in m³ T = turnover time, expressed in hours	

Examples of guidelines for maximum turnover times are six hours (Great Lakes, 1996:22) or four to eight hours (European Union of Swimming Pool and Spa Associations, 2010:2).

10.2.4. Gutters, skimmers and bottom outlets

There are two ways to remove the surface water from the pool:

- A continuous channel around the perimeter at the top of the pool wall, at the level of the water in the pool and surrounding bather area. This is known as an overflow pool.
- Several outlets built into the top of the pool wall. The water level of the pool will be approximately 150–200 mm from the top of the pool wall and halfway up the outlets. These outlets are known as surface water skimmer outlets and the pools as skimmer pools

How much of the recycle flow should come from the surface overflow and bottom outlets respectively? The following guidelines are provided (EUSA, 2010):

- For skimmer pools, the water removed from the pool via the skimmers should be 70% of the circulation rate, the remainder being removed from the low-level outlets.
- For overflow pools, the water taken from the overflow gutters should be 75% of the total

---

Example 10.2

Calculate the surface and bottom outlet flow rates for a skimmer pool with a volume of 230 m^3 and turnover time of 4.5 hours.

Total recirculation rate = 230 / 4.5 = 51.1 m^3/h

Percentage of flow through bottom outlet = 30%

Flow through bottom outlets = 0.30 x 51.1 = 15.3 m^3/h

Flow taken from the surface = 51.1 – 15.3 = 35.8 m^3/h

10.2.5. Bathing load

How do we determine the number of bathers that can be safely accommodated in a swimming pool? Table 10.1 provides a comparison between two different authorities.

Table 10.1 Guidelines for required pool area per bather

	Canada [a]	USA [b]
depth <1.5m	0.93 m^2	1.39 m^2
1.5m< depth	2.5 m^2	2.32 m^2

[a] British Columbia (2003); [b] Great Lakes (1996)

The different areas at different depths in Table 10.1 are necessary because bathers can stand upright in shallow water without moving. In deeper the water, each bather needs more area to swim or float.

Example 10.3

Use the guidelines in Table 10.1 and the pool dimensions in the diagram below to calculate the maximum recommended bathing load.

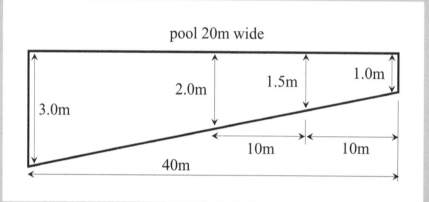

Pool area with depth between 1 m and 1.5 m = 10 x 20 = 200 m²

Pool area with depth more than 1.5 m = 30 x 20 = 600 m²

Bathing load Canada = 200 / 0.93 + 600 / 2.5 = 215 + 240 = 455 bathers

Bathing load USA = 200 / 1.39 + 600 / 2.32 = 144 + 259 = 403 bathers

All pools should have a deck or paved strip with minimum width of 1.5 m all around the pool. However, when more recreational area is provided, some of the bathers might be occupied with sport or picnicking. A pool area surrounded by large recreational space can therefore accommodate more visitors. In this case, the bathing load can be increased as follows (Great Lakes 1996):

- One extra bather for every 4.6 m² of pool deck in excess of the minimum 1.5 m required.
- One extra bather for every 9.3 m² of picnic and play area within the pool enclosure.

Example 10.4

A pool with length of 13 m and width 6 m is surrounded by a deck 3 m wide. How many additional bathers may be allowed according to the USA standard? (The minimum deck width is 1.5 m.)

Minimum deck area = (13 + 3) x (6 + 3) – (13 x 6) = 144 – 78 = 66 m^2

Actual deck area = (13 + 6) x (6 + 6) – (13 x 6) = 228 – 78 = 150 m^2

Deck area more than the minimum required = 150 – 66 = 84 m^2

Additional bathers allowed = 84 / 4.6 = 18 bathers

Example 10.5

If a picnic area of 1 200 m^2 and a volleyball court is planned within a pool enclosure, how many more bathers may be allowed according to the USA standard? (Standard volleyball court dimensions = 18 m x 9 m.)

Volleyball court area = 18 x 9 = 169 m^2

Picnic area = 1 200 m^2

Total recreational area = 169 + 1200 = 1 369 m^2

Additional bathers allowed = 1369 / 9.3 = 147 bathers

10.2.6. Balancing tank

Water does not flow evenly off the surface of a pool, regardless whether skimmers or gutters are used. Bathers making waves in a pool, large numbers of bathers suddenly jumping into a pool, or bathers playfully blocking outlets for short periods cause uneven withdrawal of the water through the skimmers or gutters. It requires a balancing tank to even out the flow disturbances before the water is pumped at an even rate through the filter. (For domestic pools, the pump is connected directly to a submerged outlet and therefore "takes" the water at an even rate, obviating the need for a balancing tank.)

Water should flow over the top of the pool into the channel over the whole perimeter when the pool is in use and over 90% of the perimeter when it is not in use. To ensure even overflow, the final finish of the top edge of the pool wall must not deviate more than ± 2 mm over 25 m. It is essential that the water running into the balancing tank can flow unrestricted under gravity.

The volume of the balancing tank should be large enough to accept the maximum anticipated surges in flow, plus a margin of safety. The maximum surge is related to the surface area of the pool and it is suggested that the minimum tank volume should be taken as 40 L/m^2 of pool area (Great Lakes, 1996).

Example 10.6

Calculate the required volume of a balancing tank for a municipal pool with volume 1050 m³ and average depth of 1.8 m.

Pool area = 1050 / 1.8 = 583 m³

Surge capacity = 40 x 583 = 23 320 L

Balancing tank assumed to absorb the full surge, therefore volume ≈ 24 m³

Another approach is to size the balancing tank according to the maximum number of bathers anticipated, with a design guideline of 57 L/bather (British Columbia, 2003).

Example 10.7

Use the Canadian standard to estimate the volume of the balancing tank for the pool in Example 10.6. Assume that 40% of the pool area has a depth more than 1.5 m.

Area deeper than 1.5 m = 583 x 0.4 = 233 m²

Area shallower than 1.5 m = 583 – 233 = 350 m²

Bathing load = 233 / 2.5 + 350 / 0.93 = 469 bathers

Surge capacity required per bather = 57 L

Volume of balancing tank = 57 x 469 ≈ 27 m³

10.2.7. Filtration system

The filter size should be in proportion to the pool volume. This is ensured by sizing the filter in terms of its hydraulic loading during normal circulation. The hydraulic loading on the filter bed is calculated as the flow rate divided by the area of the filter bed. (You could also think of it as the velocity of the water flowing downward towards the sand bed during filtration.) If the filter is undersized with a small filter bed, the hydraulic loading becomes too high for efficient filtration. This is observed when little bits of debris appear in the filtered water, or when the rapid clogging of the filter demands backwashing at short intervals with excessive water loss. A maximum hydraulic loading of 10.2 L/s/m² of filter area is proposed, provided that the filter had been certified to work effectively at that rate (Great Lakes, 1996).

During backwash mode, the important variable is the backwash rate, which is in principle the same concept as hydraulic loading, but now with the water flowing upwards through the sand bed. The backwash rate should find a compromise between being high enough to fluidise the sand bed to dislodge the trapped particles, but low enough not to wash the sand media out of the filter housing. The maximum backwash rate is suggested as 10.2 L/s/m² of filter area (Great Lakes, 1996).

It is no coincidence that the maximum hydraulic loading and the maximum backwash rate have the same value. Filter designers, over many years, had taken great trouble to get these numbers to be about the same. If they are the same, then the same pump can be used for circulation and backwash, thereby eliminating the need for an extra pump with associated pipework and valves.

Example 10.8

Calculate the minimum area of the sand bed required for filtering the recirculation flow of a pool with volume 520 m³ and maximum turnover time of 4 hours, based on the maximum hydraulic loading mentioned in the text.

Recirculation rate = 520 / 4 = 130 m³/h = 36.1 L/s

Hydraulic loading = 10.2 L/s/m²

Minimum filter area = 36.1 / 10.2 = 3.54 m²

Minimum filter diameter (if a conventional round filter is used) = 2.13 m

---

The nature of centrifugal pumps (which are predominantly used for swimming pools) is such that the flow rate is dependent on the resistance offered by the filter. A clean filter, recently backwashed, offers less resistance than a fully clogged filter. The recirculation rate is therefore more through a clean filter than when the filter is dirty. To limit the swings between maximum and minimum recirculation rates, more than one filter can be installed in parallel, an option limited to large pools with large pumping installations. The hydraulic analysis and detailed design of pumps, pipes, valves, tanks, filters, and gutters are the domain of engineering design and not discussed here.

10.2.8. Backwash rate and frequency

The backwash frequency is variable, as it depends on the number of bathers, the ingress of dust and leaves owing to wind, water temperature and others. The need to initiate backwashing is signalled by either one of the following conditions, whichever occurs first:

- If the recirculation rate drops below the threshold value where the turnover time becomes longer than the minimum.
- If the time since the previous filter backwash exceeds a stipulated maximum. If filters run too long between backwashes, biological organisms can multiply within the filter; so, it is customary to limit a filter run to say four days or one week.

Once a backwash is initiated, how long should it be continued? Excessively long backwash times translate into large water losses. A backwash should be continued until the backwash water becomes clear. If the waste stream can be observed, the backwash can be terminated by eye, but it is more common to terminate the backwash with an electronic timer set after some local experience.

10.3. POOL CHEMISTRY

10.3.1. Why pool chemistry is important

Pool owners and operators are mostly more concerned with pool disinfection (the subject of Section 10.4), but it is necessary to consider the broader issue of general pool chemistry first. There are three reasons for being concerned about the pool chemistry:

- The water should neither be corrosive nor scale-forming, in other words, stable. This is an issue covered earlier in Chapter 5. To repeat, the pH should neither be too high (scale-forming) nor too low (corrosive) but should be at neutral pH_s.

- The chemical parameters should be kept within certain limits to avoid irritation or harm to bathers.

- The chemical parameters should be such to allow the maximum effect of the disinfectants added for microbiological safety. In this chapter, the discussion is limited to the use of chlorine as disinfectant. Chlorine chemistry was also discussed earlier in Chapter 6. In short, the pH should be low enough to avoid the dissociation of hypochlorous acid into less effective hypochlorite

10.3.2. Individual chemical parameters

Perfect stability of water in relation to calcium carbonate depends on having the following chemical parameters in balance:

- The pH of the water. The optimum value is about pH 7.2 with an allowable range between about pH 6.8 and pH 8.0. Outside this range, the water is likely to be either cloudy (scale-forming) or corrosive to unprotected cementitious or metallic surfaces. In addition, the pH of human tears (technically lacrymal eye fluid) is in the same range, so values outside this range will cause eye and skin irritation.

- The alkalinity of the water should ideally be between 70 and 150 mg/L, expressed as $CaCO_3$. If the pH is kept in the range mentioned in the previous bullet, practically all the alkalinity will be in the bicarbonate form. If the alkalinity is too high, upward pH drift will be noticed. If too low, the buffer capacity of the water is too low and "pH bounce" will be observed, namely, big swings in pH brought on by small influences from bathers or chemicals. It is important that an appropriate analytical method is used to measure alkalinity, especially if in the presence of cyanuric acid, to be discussed in Section 10.4.4.

- The ideal calcium hardness lies between 125 and 275 mg/L, expressed as calcium carbonate. Note that the magnesium hardness does not affect water stability; so, it is important to measure the calcium hardness only – not the total hardness.

- The total dissolved solids (TDS) should be kept below 1500 mg/L to avoid eye irritation. This limit is not applicable to saltwater pools. The TDS will always tend to creep up owing to addition of chemicals, impurities coming from bathers, dirt and other contaminants entering the pool.

For aesthetic purposes and safety, pool water should be clear. A simple check is whether a 75 mm to 150 mm disc, half white and half black and lying on the floor at the deepest point of the pool, is visible from the nearest side.

10.3.3. Manipulating chemical and physical parameters

What to do if one or more of these chemical parameters are out of the recommended ranges? One of the easiest, but also wasteful and expensive options is to drain all or part of the pool and refill it with tap water of better quality. To lower the TDS or the cyanuric acid concentration, this is unfortunately the only option.

Example 10.9

The TDS of a pool has reached an unacceptable level of 1 850 mg/L. What percentage of the pool must be replaced with tap water if the tap water has TDS of 230 mg/L and the target TDS for the pool is 750 mg/L?

Fraction of pool volume to be replaced = P

TDS balance: $750 = 230 \times P + 1850 \times (1-P)$ which leads to $P = 0.68$

Pool volume to be replaced = 0.68 or 68%

Should the calcium hardness be too low, the calcium concentration can be elevated by adding calcium chloride dihydrate. If calcium hardness is too high, there are two options. It could be lowered by dilution or replacement with tap water, using the same calculation demonstrated in the TDS example. Alternatively, the calcium problem can be mitigated by adding a sequestering agent like sodium metahexaphosphate to mask the calcium ions with a stable complex to prevent the calcium from undergoing chemical reactions.

The water pH is lowered by adding commercial pool acid, usually diluted hydrochloric acid. The water pH can be elevated by adding sodium carbonate (soda ash) or sodium hydroxide (caustic soda). The amount of pool products required cannot be so easily calculated as in the TDS example. It is better and easier to add small amounts and to leave the pool overnight. After checking the pH the next morning, the process must be repeated until the target pH is reached. It is best to add the acid at the outlet box of the pool for good dispersion of the acid in the pool.

Alkalinity can be increased by adding sodium bicarbonate. After each addition, allow two or more days before measurement to ensure complete dissolution and mixing.

The clarity of the water should improve if the proper recirculation and backwash regimes are followed. If pools become cloudy, flocculants are used to speed up the process by binding with the contaminants to form larger flocs to be trapped in the filter.

10.4. POOL DISINFECTION

Numerous disinfection options are available, of which chlorine is by far the method of choice in South Africa, which is the only one discussed in this section. For the use of lithium or bromium-based products, ozone, chlorine dioxide or ultraviolet radiation, the many excellent sources on the internet should be consulted.

10.4.1. Chlorine chemistry basics

To provide context for the rest of the chapter, a few key points are highlighted:

- Chlorine is most effective as disinfectant when it is in the form of hypochlorous acid HOCl.

- At higher pH, hypochlorous acid dissociates into hydrogen cations and hypochlorite anions, shown earlier in Figure 6.1. Hypochlorite anions only have about 100th of the disinfection power of hypochlorous acid. For the effective use of chlorine, the pH must be kept within the pH range suggested, which guarantees that almost all the chlorine remains in the hypochlorous acid form.

- Should there be ammonia (NH_3) in the water, then chlorine will react immediately with the ammonia to form monochloramine, which has only a 10th of the disinfecting power of hypochlorous acid.

- If more chlorine is added, it reacts with the monochloramine through a series of intermediate stages, to eventually get to the point where all the remaining chlorine exists as hypochlorous acid and/or hypochlorite.

- The monochloramine and intermediate products are measured as combined chlorine and the hypochlorous acid and hypochlorite as free chlorine.

- Hypochlorous acid is gradually depleted in the presence of ultraviolet light. In open pools, exposed to the sun, chlorine must be added frequently to make up for the loss. The rate of chlorine loss can be slowed down by the addition of cyanuric acid. This leads to a distinction between unstabilised chlorine and stabilised chlorine — the latter indicating that cyanuric acid is working in conjunction with chlorine. Combined chlorine, although less effective than free chlorine, is more stable and hardly affected by sunlight.

10.4.2. Different forms of chlorine

Chlorine gas is a pure form of 100% Cl_2, delivered to site in pressurised containers. Chlorine gas addition causes a drop in the water pH.

$$C_{12} + H_2O \rightarrow HOCl + HCl$$

It is a potent gas and a chlorine gas leak can cause serious injury or death, therefore requiring full-time, certified pool operators.

Calcium hypochlorite is a granular product, well known in South Africa by its tradename HTH. Its reaction, which increases the water pH, is:

$$Ca(OCl)_2 + 2H_2O \rightarrow 2HOCl + Ca^{++} + 2OH^-$$

Sodium hypochlorite (the active ingredient in liquid household bleach) dissociates as follows with a concomitant rise in water pH:

$$NaOCl + H_2O \rightarrow HOCl + Na^+ + OH^-$$

Sodium hypochlorite can also be generated by electrolysis from a solution of sodium chloride (salt) and water. When the saltwater passes through an electrolytic cell in the circulation system, the salt is converted into sodium hypochlorite:

$$NaCl + H_2O \rightarrow NaOCl + H_2$$

Hereafter, the reaction continues to eventually form hypochlorous acid. The overall process adds a base (sodium hydroxide) therefore increases the pH.

There are two approaches to achieve electrolysis:

- The salt concentration of the entire pool volume can be raised to a minimum of not less than 3000 mg/L. The electrolytic cell is placed in the recirculation system.
- The chlorine can be electrolytically generated outside the pool in a brine solution of salt at high purity and softened water. The weak hypochlorite solution generated is stored in an on-site tank. From here, the solution is dosed into the recirculation system at the required dosage.

Cyanuric acid, although not a chlorine product, is relevant here as it is used in tandem with chlorine to stabilise the hypochlorous acid. There are a few pertinent points to make:

- UV destroys unstabilised free chorine in direct sunlight — about half of the free chlorine could disappear within four hours of bright sunlight. Cyanuric acid, however, will slow down the disappearance significantly.
- For the combination of cyanuric acid and chlorine, a fixed ratio between the concentrations of free chlorine and cyanuric acid should be maintained, which had been established to be a minimum of 7.5%. Higher concentrations of cyanuric acid therefore dictate higher doses of chlorine.
- However, it had also been established that the disinfection efficiency of free chlorine is reduced in the presence of cyanuric acid, demonstrated in Figure 10.3. If the cyanuric acid reaches a concentration of 100 mg/L, the disinfection power of chlorine is reduced by more than an order of magnitude.
- There is no chemical pathway whereby cyanuric acid can leave the pool. It therefore slowly and inevitably builds up each time more is added. When it reaches an unacceptable level, the only remedy is to drain the pool partly or completely and to refill with tap water.
- One could purchase cyanuric acid and chlorine as two separate products, or buy stabilised chlorine as a blended product which contains both components. In the first case, independent control of the ratios and concentrations is possible and recommended for those with an understanding of the underlying chemistry. Most consumers are not even aware of whether they are purchasing stabilised or unstabilised chlorine products.

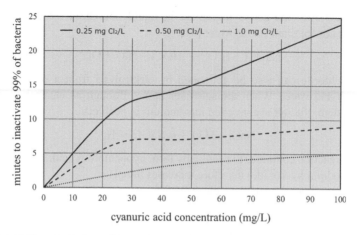

Figure 10.3 Inhibition of chlorine efficiency by cyanuric acid

10.4.3. Guidelines for unstabilised free chlorine

Guidelines for free chlorine concentrations range between 0.5 and 1.0 mg/L if the water temperature is below 30°C. To compensate for the dissociation of hypochlorous acid at higher pH, a dosage of 0.5 mg/L plus an extra 0.2 mg/L of free chlorine for every 0.2 pH unit above pH 7.2 is suggested (Great Lakes, 1996).

Example 10.10

What should the minimum unstabilised free chlorine concentration be at pH 7.8 and pH 8.5 if the Ten State Standards are followed?

    pH 7.8 is 0.6 units above pH 7.2

    Free chlorine target at pH 7.2 = 0.5 mg/L

    Increase due to higher pH = (0.6 / 0.2) x 0.2 = 0.6 mg/L

    Free chlorine target at pH 7.8 = 0.5 + 0.6 = 1.1 mg/L

    In the same way, the target at pH 8.5 = 0.5 + 1.3 = 1.8 mg/L

10.4.4. Guidelines for stabilised free chlorine

In the presence of cyanuric acid, Great Lakes (1996) suggests a free chlorine concentration of 1.0 mg/L at pH 7.2, plus an extra 0.4 mg/L for every 0.2 pH units above pH 7.2. A second approach would be to maintain the ratio of free chlorine to cyanuric acid of 7.5%. Both approaches are used in Examples 10.11 and 10.12.

Example 10.11

What should the minimum stabilised free chlorine concentration be at pH 8.2?

pH units above pH 7.2 = 1.0

Additional free chlorine = (1.0 / 0.2) x 0.4 = 2.0 mg/L

Free chlorine target = 1.0 + 2.0 = 3.0 mg/L

Example 10.12

What should the minimum stabilised free chlorine concentration be if the cyanuric acid concentration is 20 mg/L?

Free chlorine target 0.075 x 20 = 1.5 mg/L

The cyanuric acid concentration should be limited to allow the free chlorine to retain some of its disinfecting power (see Figure 10.3). Most guidelines set an absolute limit of 50 mg/L, but the operational limit should be set lower. An operating range of 15 to 30 mg/L had been suggested.

10.4.5. Guidelines for superchlorination

It was previously mentioned that free chlorine will react with any nitrogenous compounds first. After this demand is satisfied, free chlorine remains in the water for disinfection. Combined chlorine is therefore a costly obstacle to disinfection. It is unavoidable that pools will be contaminated with urine and sweat from bathers and other organic contaminants from the environment. Moreover, the presence of excessive combined chlorine can be detected by a heavy chlorinous odour and will also irritate the bathers with red eyes. For these reasons, chloramines are considered a nuisance. A maximum limit of 0.2 mg/L combined chlorine is suggested.

How can the combined chlorine concentration be reduced? The answer is to use superchlorination, which is a process of raising the free available chlorine concentration to ten times the combined chlorine concentration. This high dosage of chlorine will "shock" the pool and burn off the ammonia-based compounds. After superchlorination, free chlorine should be allowed to drop below 5 mg/L before bathers are allowed back in.

Other than reducing the combined chlorine, superchlorination also serves other purposes:

- To eliminate chlorine resistant organisms;
- To break down biofilms;
- For disinfection following a faecal event or contamination by blood or body fluid;
- To eliminate algae; and
- For disinfection following a period of high bather loading.

Superchlorination is achieved by increasing the free available chlorine in the pool far above the normal amount. Depending on the reason for superchlorination, it could be increased anywhere within the ranges of 8–50 mg/L. Unstabilised chlorine has to be used because cyanuric acid cannot be chemically reduced. When a pool is superchlorinated after a faecal accident, the high concentration of chlorine must remain in the water for a given amount of time to achieve thorough disinfection. The combination of time and concentration of chlorine is known as the CT value, covered in Chapter 7. (As a reminder, the C is the concentration of free chlorine in mg/L and T is time in hours.) A minimum CT value of 250 is commonly suggested for superchlorination.

10.4.6. Algaecides and algaestats

Algae introduced into a pool by dust or make-up water will multiply rapidly in pool water in the presence of sunlight. If not controlled, they spread rapidly with the potential to turn an entire pool completely green in as little as a day or two. Proper chlorination and water circulation in the pool basin will go a long way to prevent excessive algal growth. If not attended to frequently and quickly, some algal species will grow on the sides and bottom of the pool, in extreme cases leaving a thick, gooey biofilm. In such cases, mechanical removal by brushing and removal of the algal mass from the pool bottom is necessary — chemical means are not sufficient on its own. In extreme cases, the pool may have to be drained and the walls and floor washed down with dilute acid or hypochlorite solution. Should algae pose an ongoing problem, specialised chemicals may be considered:

- Algaecides kill algae when they have already become a problem.
- Algaestats are preventative and help to prevent the growth of algae before they become a problem.

10.5. HOW TO FIX A DIRTY POOL

As a summary to this chapter, the following troubleshooting guide is offered to address unexplained pool problems or pools after a period of neglect:

- Check the pool in circulation mode. Ensure that the pool cleaner is in good working order, that pool pipes are not punctured and that their joints are secure. Ensure that the pump is running in circulation mode and check at the inlet and outlet that water actually flows when the pump is running.
- Check the pool in backwash mode. Ensure that the changeover valves are working and that the water is flowing. Inspect the spent backwash water to make sure that the backwash water is dirty at first and that it clears after a while.
- Open the filter and make sure that it is properly filled with pool sand of correct specification.
- Measure the pool volume, the backwash rate and the circulation rate. Make the basic calculations provided in this chapter to see if the turnover time, the hydraulic loading on the filter and the backwash rate are within the recommended ranges.
- Replace part of the pool water if concentrations of cyanuric acid or total dissolved solids are too high.

- Bring pH to correct range by adding appropriate chemicals.

- Superchlorinate the pool if the combined chlorine is too high.

- Maintain the chlorine concentration at say 2 mg/L.

- Scrub the floor and sides of the pool to remove layers of biofilms and sediment accumulated on the floor. During this initial cleaning, backwash frequently. Thereafter, brush the sides every day until the pool is up to standard.

- Adjust pH and free chlorine to guideline values. The pool should return to normal within two to three days.

10.6. REFERENCES

British Columbia Health Protection Branch. 2014. *Guidelines for Pool Design Version 2*. Vancouver: Ministry of Health. https://www2.gov.bc.ca/assets/gov/environment/air-land-water/pool_design_guidelines_jan_2014_final.pdf

European Union of Swimming Pool and Spa Associations. 2010. *Circulation Systems for Domestic Swimming Pools*. Brussels: European Union of Swimming Pool and Spa Associations. https://www.eusaswim.eu/wp-content/uploads/2016/02/Paper-on-circulation-systems.pdf

Great Lakes – Upper Mississippi River Board of Public Health and Environmental Managers. 1996. *Recommended Standards for Swimming Pool Design and Operation Policies for the Review and Approval of Plans and Specifications for Public Pools.* https://docplayer.net/350594-Recommended-standards-for-swimming-pool-design-and-operation-policies-for-the-review-and-approval-of-plans-and-specifications-for-public-pools.html

Rancis, Nick. 2016 "Cyanuric Acid – Friend or Foe?" Pool + Spa, 5(1): 42-43.

World Health Organisation. 2006. *Guidelines for Safe Recreational Water Environments Volume 2: Swimming Pools and Similar Environments*. Geneva: World Health Organisation. http://apps.who.int/iris/bitstream/10665/43336/1/9241546808_eng.pdf

PART IV

BULK WATER SUPPLY

PART IV

HYDROLOGY

11.1. INTRODUCTION

The final two chapters relate to the broader issue of the availability of water which determines the water security of cities and towns. Bulk water supply may seem unrelated to the duties of the municipal chemist but is of overriding importance as demonstrated by the 2018 Cape Town "Day Zero" crisis. Municipal chemists are part of the senior corps of water managers at South African cities and towns, having to deal with these issues at one time or another. The material is spread over two chapters. Chapter 11 deals with water resources in nature while Chapter 12 will address the consequences of concentrating freshwater in large storage dams.

The field of hydrology covers a vast and complicated domain. It is not the aim of this chapter to provide methods or procedures for performing hydrological calculations, but to alert the reader to the many different aspects and applications of this important discipline.

Hydrology studies the natural movement of water through the hydrological cycle – from evaporation through cloud formation, precipitation, surface runoff all the way back to the ocean, infiltration into the soil, groundwater movement and abstraction, and so forth. The first role of hydrology is to produce estimates of the quantity of water in its different forms, which is the focus of this summary. Water quality modelling of the hydrological cycle depends both on hydrological inputs and a solid knowledge of environmental chemistry, which is a specialised discipline and not addressed here.

11.2. THE GLOBAL WATER CYCLE

The natural occurrence of water on Planet Earth is illustrated in Figure 11.1. Only 0.5% of our total water reserve is available as unfrozen freshwater, of which the largest part, by far, is below ground. Freshwater stored in reservoirs, which is the primary source of freshwater for South Africa, makes up a miniscule fraction of the total water volume on earth.

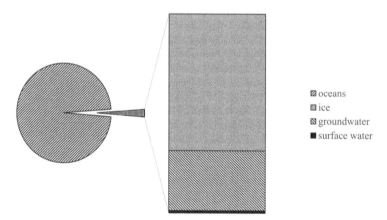

oceans
ice
groundwater
surface water

Figure 11.1 Distribution of water on earth

How much freshwater is there for every person on earth? At the moment, the average is between 5000 and 6 000 m^3 per person per year. This average has been declining sharply in the last few decades because the same amount of water being shared by an increasing number of people. Figure 11.2 shows how this volume changed since 1962. With unbridled population growth, freshwater will certainly become a constraint for human habitation.

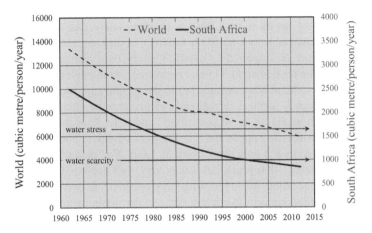

Figure 11.2 Worldwide average of available water per person per year since 1962 (World Bank N.d.)

There are large spatial variations in the distribution of freshwater as well as in the population density. If we repeat the calculation for available water per person per year for individual countries, a much more alarming picture emerges. When the available water drops to 1700 m³/year, the country is deemed to experience water stress; below 1 000 m³/year, is deemed to suffer water scarcity. Sub-Saharan Africa is the most water-stressed region in the world. The dire position of South Africa is shown in Figure 11.2.

11.3. PROBABILITY AND STATISTICS

At the outset, it should be appreciated that the movement of water within the hydrological cycle is a deeply stochastic process. While the movement of planets and the changes of seasons are predictable and repetitive, the occurrences and magnitude of floods and droughts, for example, are not. There are simply too many other factors at work. Hydrology is therefore primarily an empirical study area, where observations are made first, with explanations and mathematical models following later.

All stochastic phenomena follow a statistical distribution. Statistical distributions are such that more data points are clustered around the average, with the data points getting scarcer as one moves away from the mean. However, the main application of hydrology is to make estimates out on the wings of the distribution, where we are experiencing times of extreme drought or disastrous floods – exactly there where the data points are few and often less reliable. Having more data of acceptable quality is, therefore, indispensable for the hydrologist.

It follows that there are two main branches of hydrology, namely, flood hydrology which helps us to predict water levels during times of extreme floods, and water resource analysis which helps us to develop enough water supplies for times of water scarcity. Roughly speaking, flood hydrology concerns itself with relatively short time periods of hours to weeks, while water resource analysis considers longer time scales, measured in years to decades.

Engineering structures are designed for a specified probability of failure — scientists will appreciate that there can be no such thing as absolute safety. Our tolerance for failure depends on the anticipated consequence of the failure. The drainage structures of suburban roads, for example, may be designed to overflow every two years on average, but on national roads, it may be pushed up to 50 years on average. In the first case, it may only be a short inconvenience to local homeowners, but in the second case the national economy, the provision of emergency services or national defence may be affected. It is important to interpret terms such as the "1 in 50 year flood" correctly. This does NOT mean that the flood or drought will only happen once in the next 50 years, but that the flood will be exceeded, on average over a long time, every 50 years. For example, we might expect 20 such floods in the next 1 000 years, but it may very well happen that two such floods may follow in short succession, analogous to getting two sixes in two throws of a dice. It is a difficult but important concept to communicate to the public who lacks the scientific background to grasp this essential point.

Without getting into the details, it should also be noted that many different statistical distributions are used for the description of hydrological data. There are two general difficulties in finding a distribution that fits the data best. The first is to account for the fact that many hydrological values cannot be less than zero, such as river flow. This needs a distribution that does not allow any values less than zero, which can most simply be achieved by using the logarithms of the data to fit, for example, a log-normal distribution. The second difficulty is to assign probabilities to measured values. If we have data for 100 years, how do we assign a probability to the highest value, the second highest value, and so forth? Different mathematical models had been proposed and used.

11.4. PRECIPITATION

Precipitation is a general term which includes rain, snow and frost. In the South African context, with its warm climate, snow is a very minor factor and ignored here. Keep in mind that snow is a major cause of floods in colder climates. A major cause of floods is the accumulation of large banks of snow during the winter months, followed by a hot snap in early spring, leading to large and sudden release of snowmelt.

Coordinated rainfall measurements are made at many locations throughout South Africa. Most are made on farms, supplemented by numerous sophisticated weather stations with automatic and continuous recording of rainfall intensity, typically found at airports, dams and national parks. All the data are collected, verified and published by the South African Weather Service, which makes the data available to academics on request.

A fundamental, intuitive property of rainfall is that the intensity of rainfall depends on the duration of the rainfall event. We all know that a very hard downpour during a thunderstorm is normally limited to a few minutes. On the other hand, rainfall that persists for two or three days is normally soft with much lower intensity. In its simplest form, Equation 11.1 captures this concept:

$$D = A. T^b$$	Equation 11.1
D = total depth of precipitation in mm T = rainfall duration in hours A = constant b = constant	

The maximum depth of precipitation for the world is roughly described by A = 300 and b = 0.50. South Africa maxima are estimated with values of 200 and 0.35; for the United Kingdom by 100 and 0.43. Why are there such large differences? This has to do with different types of rainfall mechanisms in different parts of the world. The rain could come from the lifting of moist air by topography (orographic rain), heating of low-lying air by the sun (convective rain or thunderstorms), movement of warm and cold fronts (frontal rain), cyclonic action or combinations of the above.

Example 11.1

Calculate the maximum point rainfall anticipated for storm duration of 1 hour and 1 day in South Africa and the world respectively, using Equation 11.1. (Note that there are more refined methods to perform this calculation. The example only illustrates the large differences owing to location and duration.)

South Africa 1 hour: $D = 200 \times 1^{0.35} = 200$ mm

World 1 hour: $D = 300 \times 1^{0.50} = 300$ mm

South Africa 1 day: $D = 200 \times 24^{0.35} = 608$ mm

World 1 day: $D = 300 \times 24^{0.50} = 1470$ mm

For engineering design, as previously pointed out, some probability of failure is tolerated. Maximum anticipated values may therefore be reduced in accordance with a tolerable probability of failure. Table 11.1 provides a reduction factor, assuming that the observed maximum corresponds roughly with 1:100 year probability.

Table 11.1 Reduction factor to convert from 1:100 year rainfall to other return periods [a]

Return Period	Reduction Factor
2 years	0.29
5 years	0.40
10 years	0.51
20 years	0.63
50 years	0.81
100 years	1.00

[a] From Haarhoff and Cassa (2009:16)

The data presented are for point rainfall, which only has direct relevance to very small catchments, such as the roof of a building. For larger catchments, we are interested in areal rainfall, which is less than the maximum point rainfall within the area. It is, in other words, unlikely that a thunderstorm will cover all of Johannesburg at once – only certain areas will receive the brunt of the rain. To get the average over all of Johannesburg, we must apply an areal reduction factor – the larger the catchment area, the lower the areal reduction factor. The areal factor is obtained from an analysis of the topography and area of the catchment, not covered here.

Example 11.2

By how much can the 1:100 year point rainfall in South Africa be reduced if we require the 1:20 year rainfall over an area where the areal reduction factor is 0.72?

Reduction factor for probability of 1:20 years = 0.63

Areal reduction factor = 0.72

Overall reduction factor = 0.63 x 0.72 = 0.45

There are currently about 1 000 rainfall gauging stations in the national network of South Africa. It is concerning that this number is falling rapidly owing to improper management of the network, shown in Figure 11.3.

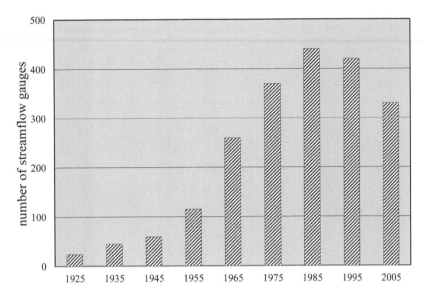

Figure 11.3 Number of registered South African rainfall gauges since 1925 (Pitman 2011)

11.5. RUNOFF

11.5.1. Direct measurement

Hydrologists use different methods to measure natural runoff in rivers and streams. Normally, a weir is built across the river with an overflow structure built to precise dimensions with a fixed relationship between flow depth and flow rate. By continuously recording the water depth, the flow rate is calculated with a high degree of precision. Using the same principle, the flow rate at all dams are also recorded during periods of overflow. Flow gauging stations are expensive to build and difficult to maintain. They are prone to vandalism and subject to being washed away or damaged during floods, which are the exact times when the most critical measurements are to be made. Sadly, the once extensive network of river flow gauging stations in South Africa is now being neglected with a drastic decline in the number of active stations, shown in Figure 11.4.

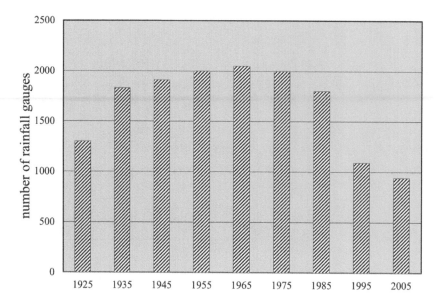

Figure 11.4 Number of registered South African runoff gauges since 1925 (Pitman, 2011)

11.5.2. Converting rainfall to streamflow

There is an obvious correlation between rainfall and runoff. If we have an accurate, rational relationship, the rainfall data, which is much easier to get, can be used to model the runoff to supplement the scarce and expensive runoff measurements. Understandably, hydrologists have been striving to find better and improved relationships linking rainfall and runoff, going back at least 150 years.

The amount of water that falls onto a catchment is readily obtained by multiplying the average areal rainfall with the area of the catchment. However, which fraction of the water falling on the surface makes it as surface runoff to the bottom end of the catchment? That depends on the catchment itself:

- The slope, surface and vegetation of the catchment. A flattish, sandy catchment will allow a substantial part of the water to infiltrate the soil while water falling on a surfaced parking lot will mostly run off before it will infiltrate.

- The same catchment could behave quite differently at different times. After a period of drought, much less water will run off owing to dry soils getting saturated, little dams and hollows being filled, and other. But a storm soon after previous rainfall will cause much greater runoff.

There are other difficulties:

- What rainfall duration should be used for estimating the runoff? This should be the same as the time of concentration, defined as the time it takes for a drop of water falling at the furthest end of the catchment to reach the point of measurement. This is estimated from the catchment properties and topography.

- What is the return period of the runoff? Bear in mind that the same storm on a "dry" catchment will cause a flood with lower return period than on a "wet" catchment.

The advent of almost unlimited computing power allows for more sophisticated hydrological modelling than before. Catchments can now be disaggregated into many smaller sub-catchments and the hydraulic behaviour of the water flow can be incorporated. Urban catchments, for example, can now be split into roofs, grassed areas, pavements, storm water pipes and more, with more detailed models directed at different components.

11.6. EVAPORATION

Evaporation is not important for flood hydrology, as the little bit of evaporation which takes place during the few hours or days of a flood is of no consequence. For water resource analysis, however, it is most important as a significant percentage of water in storage is lost owing to evaporation in our dry and hot climate.

Evaporation is measured directly in a shallow pan of 1 m² (square or circular) set up next to a rain gauge. The water level is measured every day and the tank topped up when required. If it rained during the preceding day, the measured evaporation is corrected by the rain gauge measurement. Evaporation pans are included among the instruments at more sophisticated weather stations.

Can the water loss from a shallow pan be assumed to be the same as from a large body of water? There are two factors at work here. A shallow pan does not have the same free flow of wind over its water surface, which tends to underestimate the actual evaporation. On the other hand, the water in the shallow pan warms up and evaporates faster, which tends to overestimate the actual evaporation. It turns out that the pan evaporation comes close enough to the actual evaporation for practical use.

11.7. HYDROLOGICAL REGIONALISATION

South Africa is a large country with great topographical, geological and climatological variation. The typical patterns of rainfall, runoff and evaporation are therefore also very different. For systematic study, the country is regionalised into hydrological zones at four successive levels. The primary drainage zones are at the highest level, corresponding to the main river basins. There are 22 such primary draining zones, labelled from A to X (there are no zones I or O to avoid confusion). The primary drainage zones are next subdivided into secondary zones, labelled with a number following the primary letter, for example A1, A2, etc. Next, secondary zone A1 is subdivided into tertiary catchments indicated by a further number; for example, A11, A12, etc. The final step is to divide the tertiary catchments into quaternary catchments, indicated by a final letter such as A11A, A11B, etc.

The quaternary catchments, of which there are almost 2 000, are the basic building blocks of South African hydrology. Whatever or wherever the problem, they make it easy to delineate the upstream drainage area and to identify which quaternary catchments are included. For each quaternary catchment, a mass of valuable hydrological information had been systematically assembled and tabulated. Without leaving one's desk, a great deal can be learned about the catchment from readily available publications.

As a final demonstration, a person sitting in Auckland Park, Johannesburg, can easily and quickly look up the following information to gather a rapid hydrological perspective:

- Auckland Park is situated in quaternary catchment C22B.

- It therefore is part of primary drainage zone C, which is divided into nine secondary catchments. Zone C is home to 405 rainfall stations, 65 evaporation stations, 54 runoff gauges and 312 registered dams. It has an area of 196 293 km² with mean annual precipitation of 571 mm.

- Secondary catchment C2 has an area of 35 808 km² with mean annual precipitation of 590 mm. There are nine runoff gauges and 61 registered dams.

- Tertiary catchment C22 has an area of 5 110 km² with mean annual precipitation of 657 mm. There are 12 registered dams.

- Quaternary catchment C22B has an area of 392 km² with mean annual precipitation of 691 mm.

11.8. REFERENCES

Haarhoff, J., and Cassa, A. 2009 *Introduction to Flood Hydrology.* Cape Town: Juta and Company.

Pitman, WV (2011) Overview of Water Resource Assessment in South Africa: Current State and Future Challenges. *Water SA* **37** 5, January 2011.

World Bank. N.d. Renewable internal freshwater resources per capita (cubic metres). World Bank / CC BY 4.0 / data.worldbank.org) https://data.worldbank.org/indicator/ER.H2O.INTR.PC

STORAGE DAMS

12.1. INTRODUCTION

In South Africa, dams are the backbone of municipal water supply. In an idealised world, if river flow could be perfectly constant and the demand for water could stay perfectly even, there would be no need for storage. But South Africa is known for strongly seasonal rainfall, severe droughts and water demand dominated by agriculture (more than 60%) with its uneven water demand. To balance the resulting mismatch between available water and water demand, storage is required – more of it.

Dams, however, come at a price and with some serious disadvantages. For many years they were unquestioningly and enthusiastically endorsed as primary instruments for economic development, but there is now much more caution when new dam projects are considered. This chapter summarises some of the critical issues to be carefully considered when more storage is contemplated.

12.2. DAM CLASSIFICATION

12.2.1. Size

Dams are potentially the most dangerous structures built by humans. The potential energy commanded by millions of tonnes of water impounded to great height behind a dam wall is enormous. In the event of failure, the rapid release of this energy in a short time will mostly have catastrophic consequences. As dam walls are increased in height, holding back larger volumes of impounded water with more potential energy, the consequences deepen. Engineers have therefore introduced a "large dam" category for those dams posing exceptionally large risks. By international consensus, any dam having a wall higher than 15 m is considered to be a large dam. For wall heights between 5m and 15m, it may still qualify as a large dam if the crest length is more than 500 m; or the storage capacity is more than one million cubic metre; or the spillway capacity is more than 2 000 m^3/s.

Worldwide, a total of about 58 000 large dams are registered by the International Commission on Large Dams (ICOLD 2020) – about one large dam for every 120 000 people on earth. In South Africa, there were 1 112 large dams in 2018 – about one large dam for every 50 000 inhabitants. Some of the largest dams in Southern Africa in terms of storage capacity are shown in Table 12.1.

Table 12.1 Seven large dams in Southern Africa

Name	Completion	Wall Height (in m)	Surface Area (km²)	Capacity (million m³)
Gariep	1971	88	370	5 340
Vanderkloof	1977	108	133	3 188
Sterkfontein	1977	93	70	2 656
Vaal	1985	63	320	2 610
Pongolapoort	1973	89	133	2 446
Katse (Lesotho)	1996	185	36	1 950
Mohale (Lesotho)	2004	145	22	938

12.2.2. Risk

South Africa's Department of Water Affairs has a dedicated Dam Safety Office with the mission "to promote the safety of new and existing dams with a safety risk so as to reduce the potential harm to the public" (Dam Safety Regulation, 2015). In 2015, it had 5 102 dams on its register. Based on regular inspections, the Dam Safety Office classifies the risk associated with each dam from Category I (least risk) to Category III (highest risk) because of its wall height and losses (lives as well as economic) if it should fail. The 5 102 registered dams are distributed over the different categories as shown in Table 13.2.

Table 12.2. Risk classification system used by the Dam Safety Office (number of dams) [a]

	Low Losses	Significant Losses	High Losses
Height < 12m	Category I (2820)	Category II (972)	Category III (44)
12m < Height < 30m	Category II (298)	Category II (599)	Category III (128)
Height > 30m	Category III (0)	Category III (21)	Category III (147)

[a] Dam Safety Regulation 2015

The dams in Categories II and III must undergo mandatory dam safety evaluations every five to ten years. The institutional capacity to perform these in-depth evaluations is currently limited and the evaluations are falling behind. This is partly owing to vacancies at the Dam Safety Office but also to the small and aging pool of engineers qualified for the task. In 2015, there were only 92 active Approved Professional Persons (APPs) left – 31 younger than 60, 33 between 60 and 70, and 28 older than 70.

12.2.3. Structure

Structurally, there are three basic types of dams:

- Gravity dams have a triangular cross-section with a broad base, tapering to the top. It derives its resistance to overturning by its sheer mass.
- Buttress dams arrest the overturning moment by a series of perpendicular, or buttress walls.
- Arch dams transfer the force of the water pressure to specific points where strong foundations are provided. These dams are best suited to narrow gorges where the forces are concentrated on foundations keyed into the sides of the embankments.

Figures 12.1 to 12.4 provide examples of the different dam types.

Figure 12.1: The arched wall of Gariep Dam showing hydropower station
(©Ronnie McKenzie, with permission)

Figure 12.2: The earth wall of Mohale Dam, an important part of the Lesotho Highlands
Water Scheme (©David Love / CC 0)

Figure 12.3: The concrete wall of Quedusizi Dam at Ladysmith, a gravity structure built for flood control

Figure 12.4: Roselend Dam in France with both arch and buttress features, built for hydropower. (Barrage de Roselend by versgui / CC BY-SA 3.0 / https://en.wikipedia.org/wiki/ Roselend_Dam#/media/File:Barrage_de_Roselend_2.jpg)

12.2.4. Material

Dam walls are characterised by their materials of construction:

- Concrete, which is either conventionally cast in sections or blocks ("concrete dams"), or in layers of concrete compacted by rollers ("rollcrete dams").
- Soil, which is carefully selected and compacted to provide a stable wall ("earth dams").
- Crushed rock, graded and stacked ("rock fill dams").

Soil and crushed rock are not watertight and require special precautions for waterproofing. This is done by include a clay core in the centre of the wall or placing a membrane or liner somewhere in the wall structure to provide an impermeable barrier.

12.2.5. Primary purpose

Dams are also classified according to their main purpose such as water supply, hydropower, etc. Figure 12.5 indicates the distribution of the large dams worldwide in terms of their main purpose.

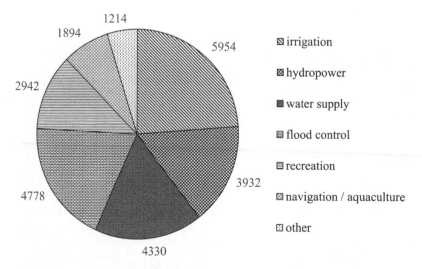

Figure 12.5: Primary purposes of registered large dams in 2018 (ICOLD 2018)

12.3. BATHYMETRY

Bathymetry has to do with the shape of the underwater world of lakes and oceans. For hydrologists, bathymetry more specifically describes the shape of the water volumes stored in our freshwater supply dams. It is intuitively clear that a shallow water body with large surface area will encourage higher evaporation losses than the same volume of water stored in a deep, narrow basin with small surface area. When engineers identify and compare different sites for the construction of new dams, bathymetry plays an important role.

The storage in our dams is usually reported as a percentage of the maximum storage volume, for example "65% full". We get this value from measuring the level of the water in the dam. To be able to convert the water level to a stored volume, we need to know the bathymetry of the dam. An example of the bathymetric relationships between water level and water volume for Klipvoor and Roodeplaat dams to the north and northwest of Pretoria is shown in Figure 12.6. These two dams have almost the same storage capacity, but quite different bathymetry. Klipvoor Dam is a shallow dam covering a large area, while Roodeplaat Dam is the opposite.

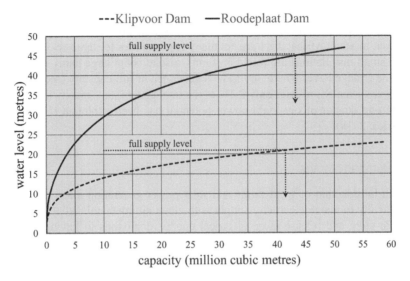

Figure 12.6: Bathymetric curves for the Klipvoor and Roodeplaat dams

12.4. ADVANTAGES OF LARGE DAMS

12.4.1. Continuity of supply

The natural inflow into dams is erratic, swinging between dry or wet years. Moreover, water demand is not constant. Large dams provide a buffer for storing water during times of excess, releasing it when demand outstrips the natural runoff. The case of the Western Cape Province during 2018 provided a good example. Rains normally come in winter, but vineyards require water in summer. When natural runoff faltered for two consecutive years, Cape Town entered a "Day Zero" crisis which made international headlines. Were it not for the large Theewaterskloof Dam which was drawn down to a mere 10% of its full capacity before the rains resumed, Cape Town would have experienced an unprecedented disaster.

12.4.2. Flood control

The buffering ability of dams is utilised to stabilise river flow during times of storms and flash floods. If a flood enters a dam which is only partially full, the first part of the flood peak is captured in the empty storage space, resulting in a diminished flood peak downstream of the dam. The Vaal Dam, for example, is operated with the water level about 4m below overflow to protect the downstream, low-lying parts of Vereeniging from inundation. The Quedusizi Dam in the Klip River upstream of Ladysmith in Natal (see Figure 12.3) serves the same purpose. Flood control dams are also used to protect agricultural lands. The Beervlei Dam between Aberdeen and Willowmore is left empty in order to protect the fertile citrus orchards in the Gamtoos River Valley further downstream from being flooded.

12.4.3. Hydropower

In some parts of the world, the abundance of water and suitable topography are ideal for hydropower generation, such as Norway where 98% of all electricity is derived from hydropower. In South Africa, there are not many opportunities for large-scale hydropower, but many small installations are found on farms, also at some large dams such as Gariep Dam on the Orange River where 360 MW is generated at full capacity (see Figure 12.2). Hydropower currently provides only 2.3% of the total generation capacity in South Africa and about one quarter of our hydropower potential has been exploited (Hydropower in South Africa, 2018).

12.4.4. Pumped storage

The recent introduction of wind farms and solar power generation made us more aware of the difficulty to match electricity supply with electricity demand. What we need is the capacity to store and release more electrical energy, analogous to having a very large electrical battery. The storage of water can do this successfully, demonstrated at numerous locations. With the wind blowing and the sun shining, we may generate more power than we consume – in this case, the excess energy is used to pump water to a higher level. At night in winter, when we consume more electricity than generated, the same water is reversed through turbines to a lower level to generate power with conventional hydropower technology. This is the principle of pumped storage.

Pumped storage, in practical terms, requires two dams at different levels, connected by large pipes or tunnels, equipped with electro-mechanical equipment which could act both as pumps and turbines. Such installations in South Africa are found at Tugela-Vaal on the border between KwaZulu-Natal and the Free State (1 000 MW), the Palmiet installation near Cape Town (400 MW) and the new Ingula installation near Ladysmith in KwaZulu-Natal (1 333 MW).

12.4.5. Aquaculture

Aquaculture offers a promising way to feed the burgeoning world population. Figure 12.7 shows how the capture of free fish had reached a plateau about 20 years ago. The gain in fish production since then is exclusively the result of an exponential growth in aquaculture.

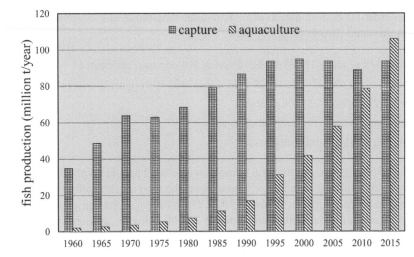

Figure 12.7: World fish capture and aquaculture production since 1960 (https://ourworldindata. org/grapher/capture-fisheries-vs-aquaculture-farmed-fish-production

The annual world production of aquaculture is roughly 25 kg per capita of which 17 kg is consumed by humans. South Africans eat relatively less fish compared to their counterparts elsewhere in the world and we consume only 7.5 kg per capita per year. Nevertheless, even if this is taken into account, our aquaculture production still lags far behind - less than 0.1 kg per capita per year. Marine aquaculture (mostly abalone, oysters, mussels and prawns) contributes about 40% of the South African total and freshwater aquaculture (trout, tilapia, catfish and carp) the rest. The Western and Eastern Cape provinces account for 95% of the country's aquaculture production. Moreover, large dams offer the opportunity to increase aquaculture significantly, an opportunity not exploited yet in South Africa (Food and Agricultural Organisation, 2018), predominantly blamed on overly stringent bureaucratic controls.

12.4.6. Recreation

Large water bodies naturally attract a variety of sportsmen and women to swim, angle, row canoes, ski or race with their power boats, and sail their yachts. In most cases, these water bodies also attract real estate investment in terms of holiday homes, hotels and resorts. All the activities attract capital investment and work opportunities and are, therefore, of national strategic value.

12.5. DISADVANTAGES OF LARGE DAMS

12.5.1. Displacement effects

Dams are large structures and, along with the associated water bodies, occupy large areas. Their immediate effect is to displace people and to flood productive agricultural land.

There are numerous case studies to demonstrate the magnitude of the problem and how it was underestimated (Bosshard, 2015; Sims, 2001):

- Ruzizi II (Rwanda) – 15 000 displaced against 200 originally anticipated.
- Sardar Sarovar (India) – 350 000 displaced against 33 000 originally anticipated.
- Kariba (Zimbabwe) – 57 000 displaced.
- Three Gorges (China) – 1 200 000 displaced.
- Merowe (Sudan) – 50 000 displaced.

Although difficult to estimate, the total number of displaced people for all dams is between 40 million to 80 million. For India and China alone, the total is estimated to be between 26 million and 56 million between 1950 and 1990. The building of the Katse Dam in Lesotho forced 2 000 people to be resettled, with another 27 000 living on its shoreline having their lives affected in one way or another.

Large dams do not only displace people, but also agriculture. The five largest South African impoundments listed in Table 12.1 had taken an area of more than 800 km² along fertile river banks out of agricultural production.

12.5.2. Ecological effects

Large dams interfere with natural phenomena in numerous ways. Rivers, before the building of dams, maintained a dynamic equilibrium with the flora and fauna on their banks. Times of low flows led to overgrowth on their banks, interspersed with flood periods which scoured the river bottoms and banks. Fish and other organisms had to migrate up and down rivers to seek the optimal ecological niches for survival. The building of large dams led to ecological fragmentation with much more even flow, less biodiversity and other unintended consequences. It is difficult to put an economic value on these effects. Legal challenges are mostly based on the possible extinction of single species, rather than being able to demonstrate the broader, more general effects which are difficult to quantify. Although some ameliorative measures can be implemented, such as fish ladders across dam walls and peak releases from time to time to simulate floods, they obviously cannot fully duplicate the natural processes.

In South Africa, some environmental concern is evident from the legal acceptance of the environment as a legitimate water user. A portion of the natural river flow is reserved to maintain some flow for the benefit of the environment (National Water Resource Strategy 2004). Provisional assessments indicate that, as a national average, about 20% of the total river flow is required as ecological reserve, which needs to remain in the rivers to maintain a healthy biophysical environment. This proportion, however, varies across the country, from about 12% in the drier parts to around 30% in the wetter areas.

12.6. IMPLEMENTATION PROBLEMS

Large dams are mega-projects which need vast amounts of money to construct. Unsurprisingly, proponents of these projects and those that are likely to benefit are anxious to see these projects go ahead. An analysis of completed projects exposed some flaws recurring in their initial feasibility studies:

- The benefits of planned projects were mostly overstated. A survey indicated that the actual hydropower realised was only 36% of what was promised, while the financial benefits from irrigation were only 60% to 85% of the projected.

- The anticipated disadvantages were understated. It took 14% to 280% longer to build the dams than planned and the cost overruns for an average of 30 dams were 342%. Dams silted up more than predicted and many more people were displaced than originally promised.

- The costs of some significant items were ignored. Earlier, no allowance was made for the decommissioning and removal of dams. More than 500 dams have now been removed in the USA alone with more targeted. The removal costs are proving to be high.

- Some items cannot be readily quantified, such as the inundation of heritage or burial sites, ecological fragmentation, and others.

During the 1970s and 1980s there was a boom in the building of big dams, with the World Bank as the major funding agency. But in the 1990s, owing to these reasons, there was growing concern and controversy on the role of large dams. The World Bank, for example, funded 26 dams per year during the period from 1970 to 1985, but during the 1990s, the funding was cut back to only four dams per year. By 2010, the international investment in large dams was still below the level of the 1980s.

These concerns prompted the World Bank and the World Conservation Union in 1998 to jointly establish a new World Commission on Dams with 12 commissioners to consider these problems and to provide guidance to planners and funders on the way forward. The first chairperson was South Africa's Minister of Water Affairs, Kader Asmal. No less than 68 concerned institutions from 36 countries worked together with the World Commission on Dams and a report was published in November 2000 with useful guidelines. Ten years later, it was evident that there was a greater awareness of the social and environmental consequences of dams, but that the guidelines were only partly adhered to. The worst culprits were among the developing nations. South Africa, after review by the World Wildlife Fund, was considered to try its best.

12.7. THE WAY FORWARD

There is no question that large dams are essential and unavoidable. At the same time, it is now crystal clear that they carry large engineering, financial and social risks. If such projects are unavoidable, they must comply with the highest ethical and professional standards.

The funding of large dam projects is shifting from the public to the private sector, creating more room for careful economic analysis and less room for scoring political points. Investors are advised to consider the following alternative options first before supporting a large dam project:

- Exploit the best practices for reducing water demand and power demand by households, industry and agriculture;
- Reduce water distribution losses;

- Consider aquifer storage as alternative where geology allows;
- Investigate direct and indirect water reuse;
- Investigate the improvement of existing dams before embarking on new ones. Norway, for example, is upgrading its hundreds of hydropower dams, large and small, by retrofitting older dams with modern, more efficient turbines to gain substantial generating capacity.

Against this background, proposals for new dams should only be considered after strategic environmental assessment to ensure that whole river basins are sustainably managed.

Once this hurdle is cleared, the critical checkpoints are (World Wildlife Federation, 2003):

- Assessment of benefits: Are the estimates of lifespan, hydropower, irrigation, tourism, recreation, fishing, etc. realistic?
- Cost recovery and dam beneficiaries: Do the charges for beneficiaries fully reflect the real costs, hidden costs and state support?
- Time and cost overruns: Is realistic allowance being made for technical problems, legal challenges and political interference?
- Displacements: Are there realistic estimates of the costs of resettling, and are there resettling agreements on resettling coasts, acceptable sites and with full involvement of all stakeholders?
- Environmental, geotechnical and social impacts: Were they considered independently without vested interests?
- Maintenance and decommissioning: Are the costs, responsibilities and decommissioning requirements clearly spelled out?
- Corruption: Will the tender process by truly open and transparent and are the safeguards against corruption matched with the country's rating on the Transparency International's Corruption Perception Index?

12.8. REFERENCES

Bosshard, Peter (2015) Dammed, Displaced and Forgotten. https://www.internationalrivers.org/blogs/227/dammed-displaced-and-forgotten

Dam Safety Regulation. 2015 *Annual Report 2014/2015*. Pretoria: Department of Water and Sanitation. http://www.dwa.gov.za/DSO/Documents/Dam%20Safety%20Regulation%20Annual%20Report%202014-15.pdf

Food and Agricultural Organisation. 2017. *World Aquaculture 2015: A Brief Overview*. Rome: Food and Agricultural Organisation. http://www.fao.org/3/a-i7546e.pdf

Food and Agricultural Organisation. 2018. *National Aquaculture Sector Overview: South Africa*. Rome: Food and Agricultural Organisation. http://www.fao.org/fishery/countrysector/naso_southafrica/en

Global Freshwater Programme. 2005. *To dam or not to dam? – Five Years on from the World Commission on Dams*. Zeist, Netherlands: World Wildlife Fund. https://d2ouvy59p0dg6k.cloudfront.net/downloads/2045.pdf

Hydropower in South Africa. At http://www.microhydropower.net/rsa/, visited July 2018.

International Commission on Large Dams (ICOLD). 2020. www.icold-cigb.net

National Water Resource Strategy. 2004. Chapter 2 – South Africa's Water Situation and Strategies to Balance Supply and Demand. Pretoria: Department of Water Affairs. http://www.dwa.gov.za/Documents/Policies/NWRS/Sep2004/pdf/Chapter2.pdf

Sims, Holly (2001) Moved, Left No Address: Dam Construction, Displacement and Issue Salience. *Public Administration and Development.* At https://onlinelibrary.wiley.com/doi/pdf/10.1002/pad.165 , visited July 2018.

Subasinghe, Rohana. 2017. *World Aquaculture 2015: A Brief Overview.* Rome: Food and Agriculture Organisation. http://www.fao.org/3/a-i7546e.pdf

World Wildlife Fund. 2003. *An Investor's Guide to Dams.* Surrey: World Wildlife Fund. https://d2ouvy59p0dg6k.cloudfront.net/downloads/investorsguidedams.pdf

INDEX